Space Nuclear Pr[illegible] and Power

Book 1

Space Nuclear Radioisotope Systems

David Buden

Space Nuclear Propulsion and Power Book 1: Space Nuclear Radioisotope Systems

Polaris Books
11111 W. 8th Ave., Unit A
Lakewood, CO 80215

www.polarisbooks.net

Manufactured in the United States of America

First Polaris Books Edition 2011

ISBN 978-0-9741443-2-0
Library of Congress Control Number: 2011933138

Cover image of Voyager 1 provided by NASA. © NASA
Book and cover design by Alyssa Piccinni

CONTENTS

FOREWORD 1
OVERVIEW 3

CHAPTER 1: RADIOISOTOPE SYSTEMS MISSIONS 11
U.S. History of Uses of Space Radioisotope Systems 12
Russian Radioisotope Applications 20
CHAPTER 2: MISSION DERIVED DESIGN REQUIREMENTS AND PLUTONIUM-238 PRODUCTION 21
Radioisotope Fuels 21
Power Conversion Systems 22
Nuclear Safety 23
New Horizons Mission 26
Plutonium-238 Production and Processing 29
CHAPTER 3: EARLIER GENERATIONS OF RADIOISOTOPE THERMOELECTRIC GENERATORS 33
Basic Building Blocks of Radioisotope Generators 33
Thermoelectric Power Conversion 37
Heat Rejection Subsystem 41
Structure and Thermal Design 43
Evolution of Radioisotope Generators 43
SNAP-3B 46
SNAP-9A 47
SNAP-19 50
SNAP-27 57
Transit-RTG 60
Multihundred Watt (MHW) Generator 65
Summary 71
CHAPTER 4: MORE RECENT RADIOISOTOPE GENERATOR SYSTEMS 73
General Purpose Heat Source-Radioisotope Thermoelectric Generator (GPHS-RTG 73
Design Configuration 73
Performance Tests 77
Safety Program 79
Mission Behavior 84
GPHS-RTG SummaryL 87
Multi-Mission Radioisotope Thermoelectric Generator (MMRTG) 88
MMRTG Design 89
MMRTG Performance 93
Environmental Impact Evaluation 93
Summary 95
CHAPTER 5: ALTERNATE POWER CONVERSION SUBSYSTEMS 97
Principles of Dynamic Thermal-to-Electric Conversion Cycles 98
Brayton Isotope Power System (BIPS) 102
Rankine Cycle Converters For Radioisotope Generators 109
Rankine Cycle Principles 109
Dynamic Isotope Power System (DIPS) Development 112
Stirling Cycles 116
Principles of Stirling Cycles 116
Advanced Stirling Radioisotope Generator (ASRG) 118
Advanced Thermoelectric Converter (ATEC) 122

Alkali Metal Thermal-To-Electric Converter (AMTEC) 124
Principles of Operation 124
AMTEC Cell Design 126
AMTEC Summary 128
Thermophotovoltaic (TPV) Generators 129
Summary 133
CHAPTER 6: RADIOISOTOPE HEATER UNITS 135
Radioisotope Heater Unit Design 137
Fuel Processing and Fabrication 140
Welding, Nondestructive Testing, And Final Assembly 141
Safety Verification Tests 142
Summary 146
CHAPTER 7: POTENTIAL FUTURE APPLICATIONS AND ISSUES 147
Potential Future Applications 147
Issues and Trends 151

NOTES 155

Foreword

Propulsion and power are defining technologies that enable man to perform more demanding space missions. In 1985, Orbit Book Company, Inc. published *Space Nuclear Power* by Joseph A. Angelo, Jr. and David Buden. Since that time, much has happened in the field of space nuclear systems. This series of three books on space nuclear power and propulsion:

Book 1: *Space Nuclear Radioisotope Systems*
Book 2: *Nuclear Thermal Propulsion Systems*
Book 3: *Space Nuclear Fission Electric Power Systems*

brings together a summary of all of the developments in space nuclear systems.

Dr. Mohamed S. El-Genk organized the Symposium on Space Nuclear Power and Propulsion in Albuquerque, New Mexico from 1983 thru 2008. Dr. El-Genk is Regents' Professor, Chemical, Nuclear, and Mechanical Engineering, and Director of the Institute of Space Nuclear Power Systems at the University of New Mexico. A debt of gratitude is owed to Dr. El-Genk for his tireless efforts in bringing together the principle investigators in the space nuclear field and publishing symposium proceedings for each of the yearly meetings. The proceedings were extremely helpful in preparing this book.

Dr. Gary L. Bennett offered invaluable insight and assistance in preparing the manuscript for this book. He worked for NASA and US Department of Energy (DoE) on advanced space power and propulsion systems. Dr. Bennett started his space nuclear career on the nuclear rocket program (NERVA) at NASA Lewis Research Center. Subsequently, he has held key positions in DoE's space radioisotope power program, including serving as the flight safety manager and Director of Safety and Nuclear Operations for radioisotope power sources used on Voyager 1 and 2 spacecraft to Jupiter, Saturn, Uranus, Neptune and beyond; Lincoln Laboratory's LES 8 and 9 communications satellites; and the Galileo mission to Jupiter. At NASA, he was the Manager of Advanced Space Power Systems in the transportation division of the Office of Advanced Concepts and Technology. He managed, among other things, fission nuclear propulsion activities.

Overview

Key elements in the exploration and exploitation of space are reliable and compact propulsion and power systems. Without adequate power and propulsion, space missions are severely limited. Nuclear powered systems are a key technology in meeting past and future propulsion and power needs. The current generation of nuclear systems can be organized in terms of radioisotope and fission nuclear reactor heat sources. Fission nuclear systems can be subdivided into power and propulsion applications. These are the bases for this series of three books:

Book 1: *Space Nuclear Radioisotope Systems*
Book 2: *Nuclear Thermal Propulsion Systems*
Book 3: *Space Nuclear Fission Electric Power Systems*

It is vital to understand past activities in developing future programs to meet the challenges that lie ahead. In order to meet the needs and interest of individual readers, each book in this series is designed to stand-alone.

Solar energy has been the cornerstone of power systems for near-Earth missions. However, the solar flux drops as the inverse square of the distance from the Sun (See Fig. 1).[1] At Jupiter, for instance, the solar flux is only 4% of that at Earth. Solar arrays become too large and weigh too much at such distances. Also, surface operations in sun starved or shadowed areas at nearer distance cannot use solar technology. Near the Sun high temperatures also limits the use of solar technology.

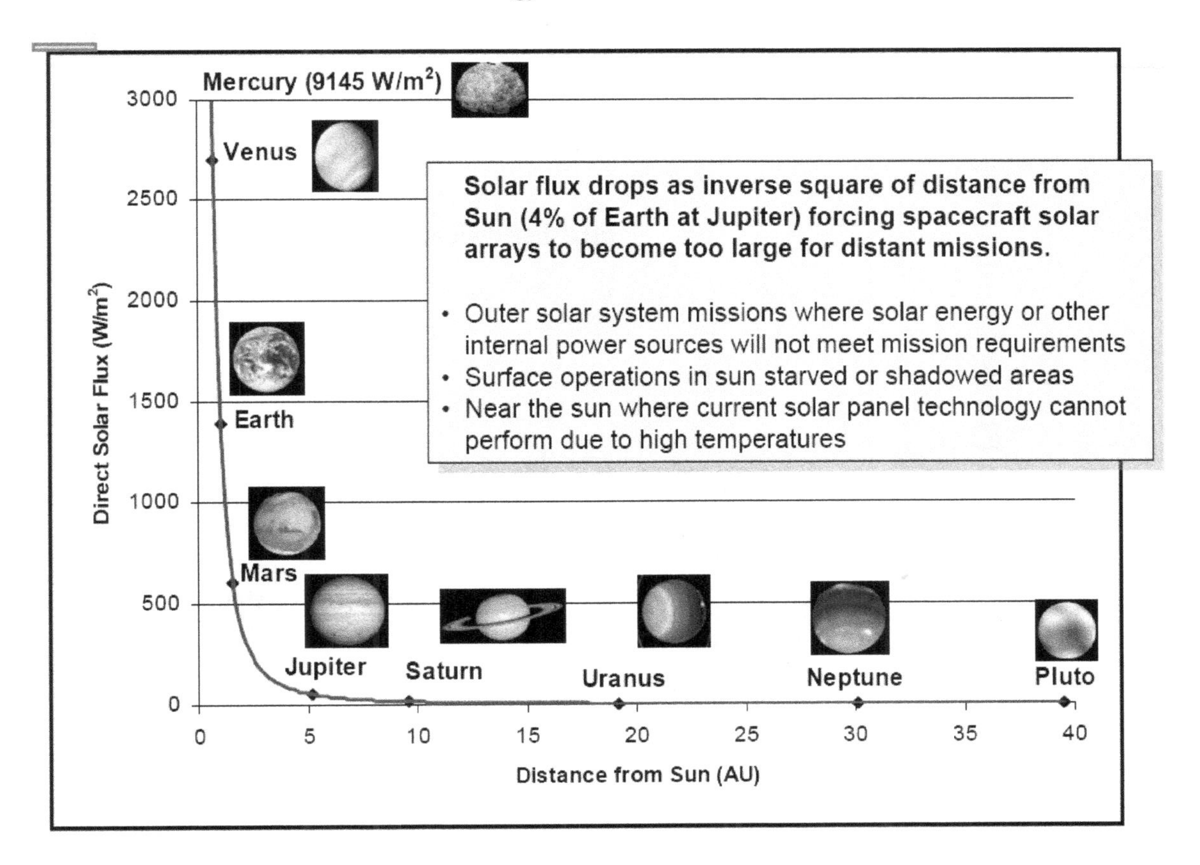

Fig. 1 Limited solar flux as distances increase from the Sun enables the need for nuclear space power systems.

To date, space propulsion has relied mainly on chemical fueled rockets. As missions with increasingly larger payloads are contemplated, the size of chemical rockets becomes unwieldy. Other more efficient propulsion systems, such as nuclear fission rockets, will be required.

Nuclear power has been used on many space missions by the United States in support of both civilian and military programs. These have taken the form of using the thermal energy from the decay of radioisotopes and converting this energy to electric power. Radioisotope power systems have proven to be highly reliable, operating for many years and in severe environments (e.g., trapped radiation belts, surface of Mars, moons of the outer planets) that make solar alternatives of limited use or unusable. Also, radioactive decay heat has been used to maintain temperatures in spacecraft at acceptable conditions for other components.

Radioisotope power systems are limited in power levels to a few kilowatts by the cost and availability of suitable radioisotope thermal heat sources. Nuclear reactors can provide tens-to-hundreds and even megawatts of power in future power systems. Extensive development works has already been performed on reactor-powered systems with the United States having flown one system in space. To meet the requirements of the more ambitious missions of the future, nuclear fission power will be a necessity.

Interest in nuclear rockets has centered on manned flights to Mars. The demands of such missions requires rockets that are several times more powerful than the chemical rockets in use today. Nuclear fission rockets have been extensively developed for this purpose. However, none of these developments have reached flight status.

The advantages of space nuclear systems can be summarized as: compact size; low to moderate mass; long operating lifetimes; operation in extremely hostile environments; operation independent of the distance from the Sun or of the orientation to the Sun; and high system reliability and autonomy.[2, 3, 4] In fact, as power requirements approach hundreds of kilowatts and megawatts, nuclear energy appears to be the only realistic power option (see Fig. 2).[5]

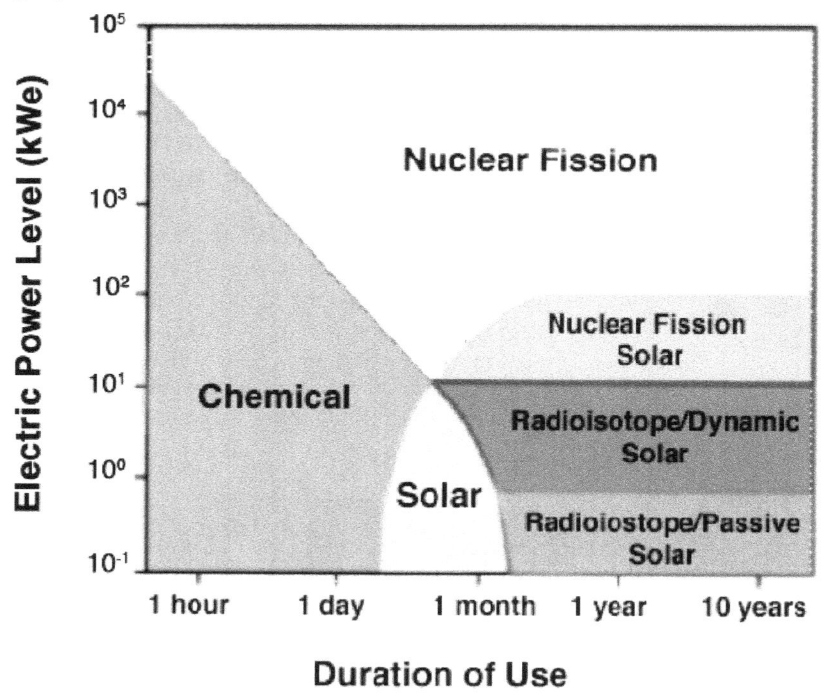

Fig. 2. Regimes of possible space power applicability.

The building blocks for space nuclear electric power system are depicted in Fig. 3. Radioisotope decay heat or the thermal energy released in nuclear fission can be converted to electrical using power generation equipment. These can take the form of either static electrical conversion elements that have no moving parts (e.g., thermoelectric or thermionic) or dynamic conversion elements (e.g., the Rankine, Brayton or Stirling cycle). The options for nuclear energy heat sources and companion power generation subsystems are summarized in Fig. 4. Radioisotope and reactor power systems are further classified in Fig. 5.[6]

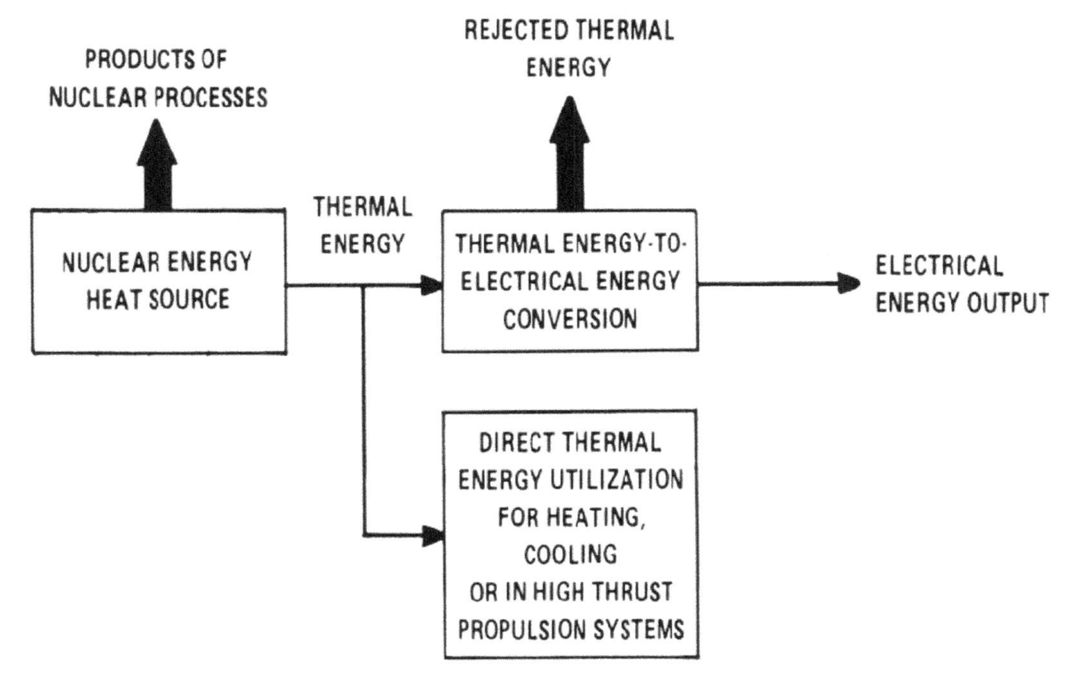

Fig. 3. Generic space nuclear power systems. ***Courtesy of Los Alamos National Laboratory.***

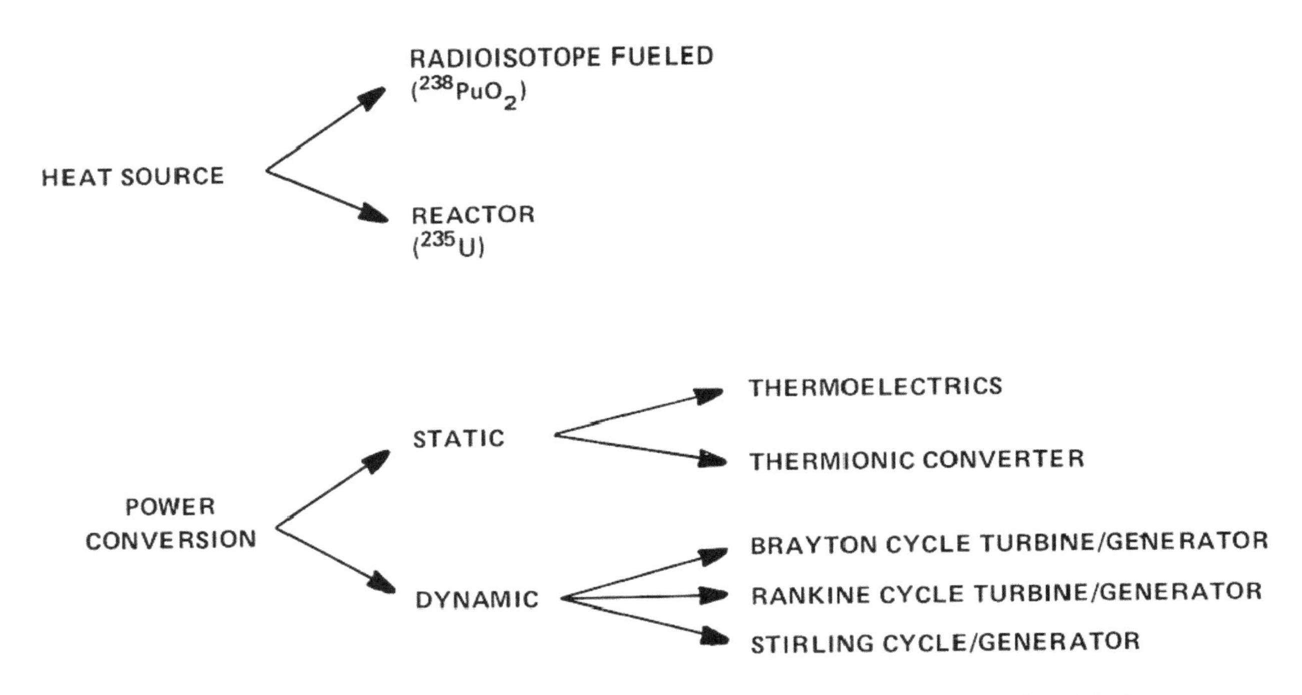

Fig. 4. Options covered in space nuclear programs. ***Courtesy of Los Alamos National Laboratory***

Nuclear Power System	Electric Power Range (Module Size)	Power Conversion
Radioisotope Thermoelectric Generator (RTG)	Up to 500 We	Static: Thermoelectric
Radioisotope Dynamic Conversion Generator	0.5 to 10 kWe	Dynamics: Brayton Rankine Stirling
Reactor Systems Heat Pipe Solid Core Thermionics	10 kWe to 1,000 kWe	Static: Thermoelectrics Thermionics Dynamics: Brayton Rankine Stirling
Reactor Systems Heat Pipe Solid Core	1 to 10 MWe	Dynamics: Brayton Rankine Stirling
Reactor Systems Solid Core Pellet Bed Fluidized Bed Gaseous Core	10 to 100 MWe	Brayton Cycle (Open Loop) Stirling MHD

Fig. 5. Classification of nuclear power system types being considered for space applications. ***Courtesy of Los Alamos National Laboratory.***

Since 1961, the United States has launched twenty-seven National Aeronautics and Space Administration (NASA) and military space systems that derived all or part of their power requirements from nuclear energy sources. These systems are summarized in Table 1; some illustrations of nuclear powered spacecraft are shown in Fig. 6. As can be seen in this table, all but one of the previous missions used plutonium-238 as the fuel in various radioisotope thermoelectric generator (RTG) systems. The SNAP-l0A was a compact nuclear fission reactor that used fully enriched uranium-235 as the fuel. The acronym SNAP stands for Systems for Nuclear Auxiliary Power, with the odd-numbered units representing radioisotope heat sources and the even-numbered units nuclear reactors.

Table 1. Summary of space nuclear power systems launched by the United States.[7, 8, 9]

Power Source	Spacecraft	Mission Type	Launch Date	Status
SNAP-3B7	Transit 4A	Navigational	6/29/1961	RTG operated for 15 years. Satellite now shutdown but operational.
SNAP-3B8	Transit 4B	Navigational	11/15/1961	RTG operated for 9 years. Satellite operated periodically after 1962 high altitude test. Last reported signal in 1971.
SNAP-9A	Transit 5-BN-1	Navigational	9/28/1963	RTG operated as planned. Non-RTG electrical problems on satellite caused satellite to fail after 9 months.

Power Source	Spacecraft	Mission Type	Launch Date	Status
SNAP-9A	Transit 5-BN-2	Navigational	12/5/1963	RTG operated for over 6 years. Satellite lost ability to navigate after 1.5 years.
SNAP-9A	Transit 5-BN-3	Navigational	4/21/1964	Mission was aborted because of launch vehicle failure. RTG burned up on re-entry as designed.
SNAP-10A REACTOR	SNAPSHOT	Experimental	4/3/1965	Successfully achieved orbit. Operated 43 days above 785 K.
SNAP-19B2	Nimbus-B-1	Meteorological	5/18/1968	Mission was aborted because of range safety destruct. RTG heat sources recovered and recycled.
SNAP-19B3	Nimbus III	Meteorological	4/14/1969	RTGs operated for over 2.5 years.
ALRH	Apollo 11	Lunar Surface	7/14/1969	Radioisotope heater units for seismic experimental package. Station was shut down 8/3/1969.
SNAP-27	Apollo 12	Lunar Surface	11/14/1969	RTG operated for about 8 years until station was shut down.
SNAP-27	Apollo 13	Lunar Surface	4/11/1970	Mission aborted on the way to the moon. RTG re-entered earth's atmosphere and landed in South Pacific Ocean. No radiation was released.
SNAP-27	Apollo 14	Lunar Surface	1/31/1971	RTG operated for over 6.5 years until station was shut down.
SNAP-27	Apollo 15	Lunar Surface	7/26/1971	RTG operated for over 6 years until station was shut down.
SNAP-19	Pioneer 10	Planetary	3/2/1972	RTGs still operating. Spacecraft successfully operated to Jupiter and is now beyond orbit of Pluto.
SNAP-27	Apollo 16	Lunar Surface	4/16/1972	RTG operated for about 5.5 years until station was shut down.
Transit-RTG	"Transit" (Triad-01-1x)	Navigational	9/2/1972	RTG still operating.
SNAP-27	Apollo 17	Lunar Surface	12/7/1972	RTG operated for almost 5 years until station was shutdown.
SNAP-19	Pioneer 11	Planetary	4/5/1973	RTGs operating. Spacecraft successfully operated to Jupiter, Saturn, and beyond.
SNAP-19	Viking I	Mars Surface	8/20/1975	RTGs operated for over 6 years until lander was shut down.
SNAP-19	Viking 2	Mars Surface	9/9/1975	RTGs operated for over 4 years until relay link was lost.
MHW-RTG	LES 8	Communications	3/14/1976	RTGs still operating.
MHW-RTG	LES 9	Communications	3/14/1976	RTGs still operating.
MHW-RTG	Voyager 2	Planetary	8/20/1977	RTGs still operating. Spacecraft successfully operated to Jupiter, Saturn, Uranus, Neptune, and beyond.
MHW-RTG	Voyager 1	Planetary	9/5/1977	RTGs still operating in 2010. Spacecraft successfully operated to Jupiter, Saturn, and in the heliosphere.

Power Source	Spacecraft	Mission Type	Launch Date	Status
GPHS-RTG	Galileo	Planetary	10/8/1989	Completed 34 orbits of Jupiter. Mission ended 21 Sept. 2003 when spacecraft was plunged into Jupiter's atmosphere.
GPHS-RTG	Ulysses	Planetary/Solar	10/6/1990	Spacecraft completed nearly three complete orbits of Sun. Operated more than 18 y, shutdown 6/30/2009.
GPHS-RTG	Cassini	Planetary	10/15/1997	RTGs still operating. Spacecraft orbiting Saturn.
GPHS-RTG	New Horizons	Planetary	1/19/2006	RTG still operating. Spacecraft en route to Pluto
MMRTG	Mars Science Laboratory	Mars Surface	Pending 2011	

The Russians have also have made extensive use of nuclear power systems in space. They launched two radioisotope systems in 1965. In the time period between 1971-1988, they launched some 35 nuclear reactor systems (see Table 2).

Table 2. USSR space power flight experience.[10]

TYPE	NAME	MISSION	NUMBER OF MISSIONS	LAUNCH DATES	STATUS	FAILURES
RTG		NAVIGATION SATELLITES	2	9/65	IN ORBIT	NONE KNOWN
RADIO-ISOTOPE HEATER UNIT		LUNAR ROVERS	4	9/69 TO 1/73	TWO SHUTDOWN ON MOON	TWO REENTRIES AFTER UPPER STAGE MALFUNCTIONS (1969)
REACTOR	RORSAT	OCEAN SURVEILLANCE	35	12/67 TO 3/88	31 SHUTDOWN AND BOOSTED TO HIGH ORBITS	- TWO LAUNCH ABORTS (1969, 1973) - TWO REENTRIES AFTER BOOST FAILURE (1977, 1982)
REACTOR	TOPAZ	OCEAN SURVEILLANCE	2	1/87 TO 10/87	SHUTDOWN AND BOOSTED TO HIGH ORBITS	NONE KNOWN

Safety has been and continues to be a key element in the development and deployment of space nuclear systems. The prevention of significant radiological risk to the Earth's population or to the terrestrial environment are guiding policies in all phases associated with space nuclear systems.[11] For radioisotope heat sources, this aerospace nuclear safety policy essentially consists of providing containment that is not prejudiced under any circumstance including launch accidents, reentry, or impact on land or water. For nuclear reactors, the safety mechanism consists of maintaining sub criticality under all conditions, normal and otherwise, in the Earth's atmosphere or on the Earth's surface. After the reactor has experienced power operation in space, the reactor will be prevented from reentering the terrestrial biosphere until the fission products and other radioactive materials no longer represent a radiological risk. Reactor operation is generally limited to orbits that have a lifetime in excess of three hundred years to support this safety approach.

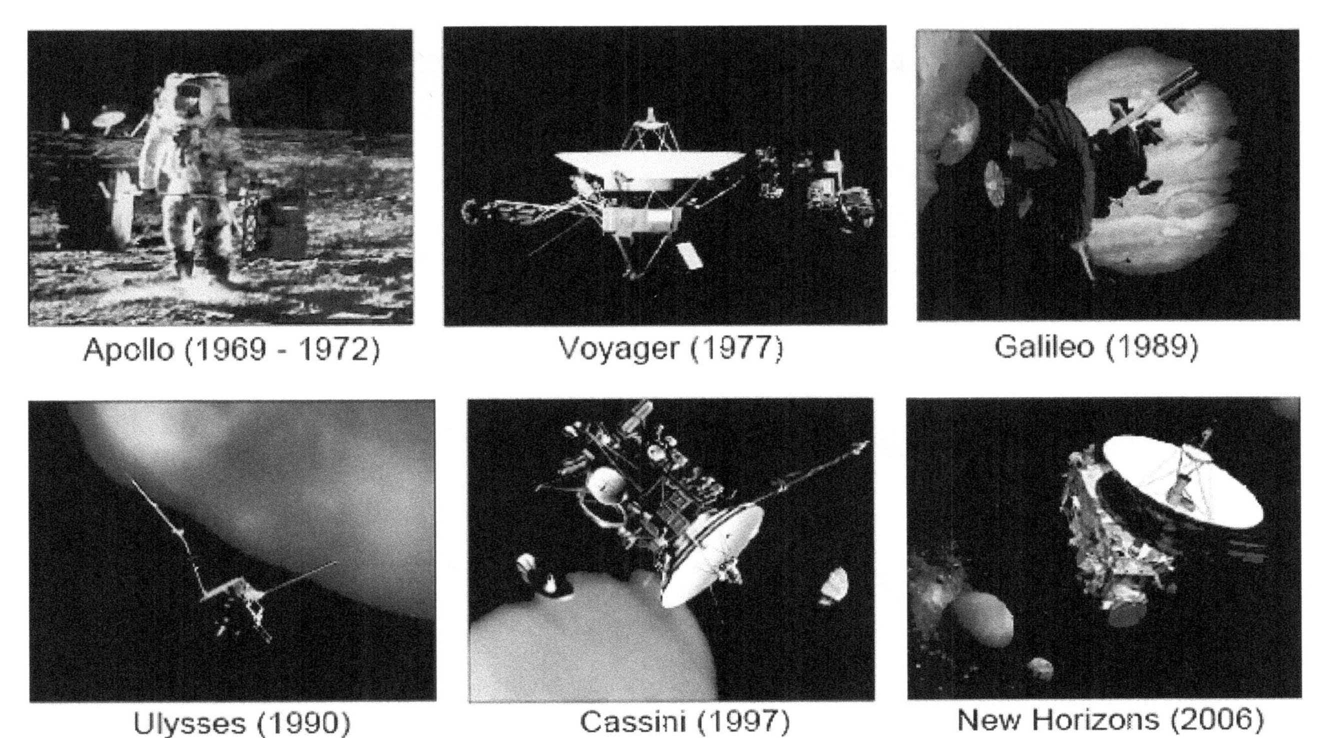

Fig. 6. Illustrations of missions using nuclear power. All missions operated far beyond their design life times. ***Courtesy of Dennis Miotia.***

Chapter 1

Radioisotope Systems Missions

For operating in severe environments, long life and reliability, radioisotope power systems have proven to be the most successful of all space power sources! Two Voyager missions (Fig. 1) launched in 1977 to study Jupiter, Saturn, Uranus, Neptune, and their satellites, rings and magnetic fields and continuing to the heliosphere region are still functioning over thirty years later. Radioisotope power systems have been used on the Moon, exploring the planets, and exiting our solar system. Their success is a tribute to the outstanding engineering, quality control and attention to details that went into the design and production of radioisotope power generation units.

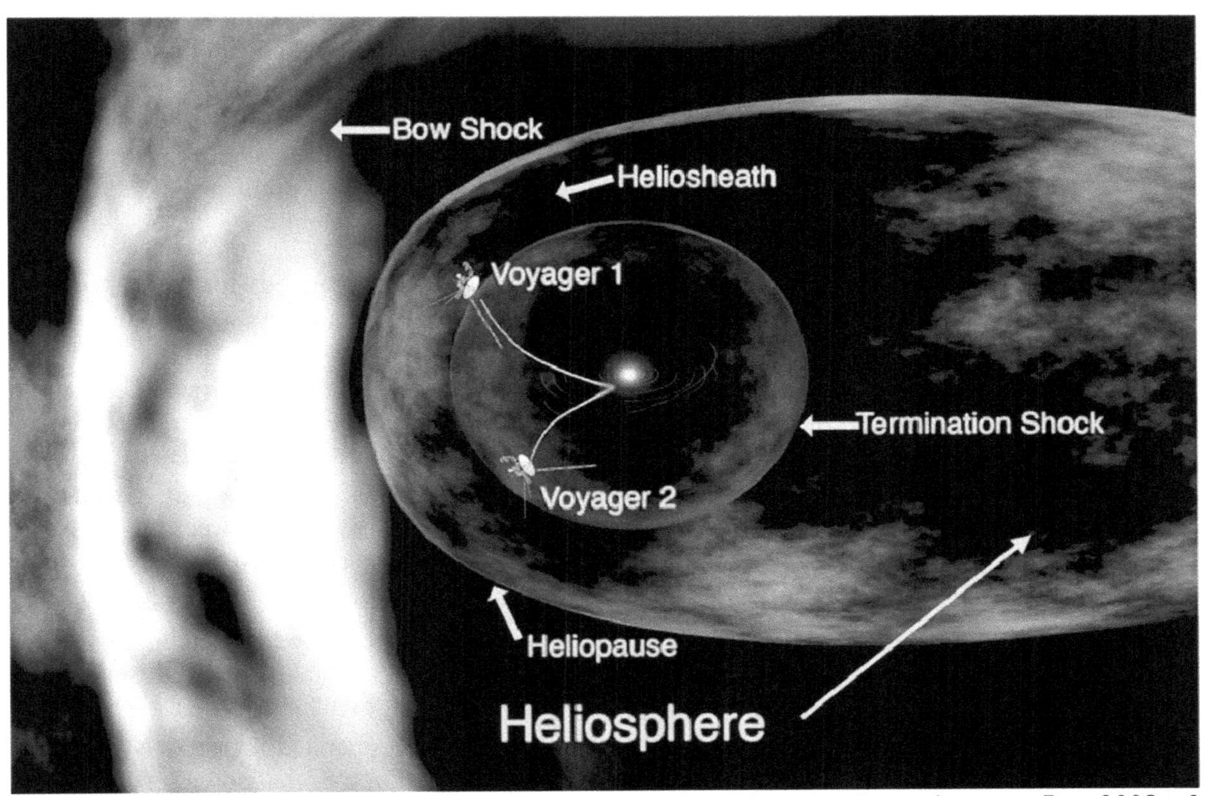

Fig. 1. Voyager 1 is the most distant human-made object in the universe. In 2008, the spacecraft was about 16 billion kilometers (approximately 10 billion miles) from the Sun. This is a region in space known as the Heliosheath where the solar wind runs up against the thin gas between the stars. The spacecraft was about 107 times as far from the Sun as is Earth. ***Image Credit: NASA/Walt Feimer***

The major limitation to the use of radioisotope power systems in future missions is their high cost. Production of Plutonium-238, the main isotope used in space power systems, has currently been discontinued because the facilities that were used in its production were deactivated. Proposed alternate production facilities have not yet been funded. There is really no good alternative to powering many of the more demanding outer planet exploration missions.

U.S. History of Uses of Space Radioisotope Systems

Since 1961, radioisotope power systems have been used by the United States to support its space program (see Table 1). The first use was on the Transit spacecraft, a navigation satellite. This was followed by radioisotopes powering weather and communication satellites; scientific stations on the Moon, robot explorer spacecraft on Mars and highly sophisticated deep space interplanetary missions to Jupiter, Saturn, and beyond. Radioisotope systems have demonstrated safe and reliable operation, the ability to operate both in hostile environments and independent of the solar flux, and long life. Several missions were aborted by launch vehicle failures--none the result of the radioisotope power system on-board. Radioisotope power systems have operated in such diverse and severe environments as the Van Allen radiation belts, the lunar surface with its temperature extremes, Martian dust storms, the asteroid belts, and in the vicinity of the frigid outer planets.[1]

Table 1. Radioisotope power systems launched by the United States.[2, 3, 4]

Power Source	Spacecraft	Mission Type	Launch Date	Status
SNAP-3B7	Transit 4A	Navigational	6/29/1961	RTG operated for 15 years. Satellite now shutdown but operational.
SNAP-3B8	Transit 4B	Navigational	11/15/1961	RTG operated for 9 years. Satellite operated periodically after 1962 high altitude test. Last reported signal in 1971.
SNAP-9A	Transit 5-BN-1	Navigational	9/28/1963	RTG operated as planned. Non-RTG electrical problems on satellite caused satellite to fail after 9 months.
SNAP-9A	Transit 5-BN-2	Navigational	12/5/1963	RTG operated for over 6 years. Satellite lost ability to navigate after 1.5 years.
SNAP-9A	Transit 5-BN-3	Navigational	4/21/1964	Mission was aborted because of launch vehicle failure. RTG burned up on re-entry as designed.
SNAP-19B2	Nimbus-B-1	Meteorological	5/18/1968	Mission was aborted because of range safety destruct. RTG heat sources recovered and recycled.
SNAP-19B3	Nimbus III	Meteorological	4/14/1969	RTGs operated for over 2.5 years.
ALRH	Apollo 11	Lunar Surface	7/14/1969	Radioisotope heater units for seismic experimental package. Station was shut down 8/3/1969.
SNAP-27	Apollo 12	Lunar Surface	11/14/1969	RTG operated for about 8 years until station was shut down.
SNAP-27	Apollo 13	Lunar Surface	4/11/1970	Mission aborted on the way to the moon. RTG re-entered earth's atmosphere and landed in South Pacific Ocean. No radiation was released.
SNAP-27	Apollo 14	Lunar Surface	1/31/1971	RTG operated for over 6.5 years until station was shut down.
SNAP-27	Apollo 15	Lunar Surface	7/26/1971	RTG operated for over 6 years until station was shut down.

Power Source	Spacecraft	Mission Type	Launch Date	Status
SNAP-19	Pioneer 10	Planetary	3/2/1972	RTGs still operating. Spacecraft successfully operated to Jupiter and is now beyond orbit of Pluto.
SNAP-27	Apollo 16	Lunar Surface	4/16/1972	RTG operated for about 5.5 years until station was shut down.
Transit-RTG	"Transit" (Triad-01-1x)	Navigational	9/2/1972	RTG still operating.
SNAP-27	Apollo 17	Lunar Surface	12/7/1972	RTG operated for almost 5 years until station was shutdown.
SNAP-19	Pioneer 11	Planetary	4/5/1973	RTGs operating. Spacecraft successfully operated to Jupiter, Saturn, and beyond.
SNAP-19	Viking I	Mars Surface	8/20/1975	RTGs operated for over 6 years until lander was shut down.
SNAP-19	Viking 2	Mars Surface	9/9/1975	RTGs operated for over 4 years until relay link was lost.
MHW-RTG	LES 8	Communications	3/14/1976	RTGs still operating.
MHW-RTG	LES 9	Communications	3/14/1976	RTGs still operating.
MHW-RTG	Voyager 2	Planetary	8/20/1977	RTGs still operating. Spacecraft successfully operated to Jupiter, Saturn, Uranus, Neptune, and beyond.
MHW-RTG	Voyager 1	Planetary	9/5/1977	RTGs still operating. Spacecraft successfully operated to Jupiter, Saturn, and beyond.
GPHS-RTG	Galileo	Planetary	10/8/1989	Completed 34 orbits of Jupiter. Mission ended 21 Sept. 2003 when spacecraft was plunged into Jupiter's atmosphere.
GPHS-RTG	Ulysses	Planetary/Solar	10/6/1990	Spacecraft completed nearly three complete orbits of Sun. Operated more than 18 y, shutdown 6/30/2009.
GPHS-RTG	Cassini	Planetary	15/10/1997	RTGs still operating. Spacecraft orbiting Saturn.
GPHS-RTG	New Horizons	Planetary	19/1/2006	RTG still operating. Spacecraft en route to Pluto.
MMRTG	Mars Science Laboratory	Mars Surface	Pending 2011	

The first radioisotope generators, SNAP-3B, provided power for the Transit 4A navigational satellite. This satellite, launched in 1961, was only 2.7 watts-electric. By 1963, the power level for the Transit navigation satellites had increased to 25 watts-electric using the SNAP-9A generator units. A new generation of Transit spacecraft was initially launched in 1972, using a 36 watt-electric RTG. Spacecraft instrumentation temperature near 293 K was maintained using "waste" heat from this radioisotope power supply. The SNAP-19 system, launched on 14 April 1969, supplied power to the Nimbus III weather satellite. Images of the

Earth's cloud cover and the investigation of the atmosphere over a wide range of the electromagnetic spectrum were produced using power from two 56-watt generators.

The Apollo Moon landings were a first to integrate nuclear systems with manned operations. These demonstrated the utility of nuclear power supplies in extraterrestrial exploration and that such systems could be safely and effectively integrated into manned missions. In an early Apollo mission, Apollo 11, a scientific experiment package (EASEP) was left on the lunar surface by the astronauts. This included two 15 watt-thermal, plutonium-238 fueled heater units to provide thermal energy. These heaters successfully kept the scientific instruments warm during the frigid lunar night (approximately 14 Earth days in duration). SNAP-27 radioisotope thermal generator systems were incorporated into subsequent Apollo missions. These generated some 63.5 watts of electric power in support of the Apollo lunar scientific experiment package (ALSEP). The radioisotope units continued to function successfully far beyond their one year design life and returned valuable scientific data long after the last Apollo astronaut had walked on the lunar surface.

Radioisotope powered Pioneer spacecraft have provided valuable scientific measurements of the giant outer planets Jupiter and Saturn. The first close-up imagery was obtained of these planets. In these truly pathfinder missions of exploration, the Pioneer spacecraft carried four modified SNAP-19 units, each generating over 40 watts-electric. The three-year mission to Jupiter included successful passage through the asteroid belt and operation in the highly intensive radiation belts surrounding Jupiter. In 1983, the Pioneer 10 spacecraft still functioning with its nuclear power supply, became the first manmade object to leave the Solar System and eventually reached the heliosphere. Pioneer 10 was successfully contacted on January 22, 2003--almost 31 years after launch.

Curiosity about the Red Planet, Mars, has interested mankind throughout the ages. Two Viking Mars Landers were launched in 1975 carrying SNAP-19 35-watt systems. The mission included nuclear powered robot landers to support the search for extraterrestrial life and to provide data about the Martian surface and atmosphere. Operation continued for four (Viking 1) to six years (Viking 2)--well after their planned two-year design life.

The multi-hundred watt (MHW) power systems were initially developed for the communication satellites named Lincoln Experimental Satellites (LES 8 and LES 9). The LES 8 and 9 spacecraft, launched in 1976, each carried two MHW units, providing some 300 watts of electric power. Further use was made of the MHW radioisotope units on the spectacular Voyager I and 2 flyby missions to Jupiter, Saturn, and beyond. Voyager 1 and 2 spacecraft, launched in 1977, each incorporated three MHW units. These spacecraft followed Pioneer 1 and 2 into the heliosphere and continue to operate successfully after 30 y.

The next generation of radioisotope power systems was the General Purpose Heat Source Radioisotope Thermoelectric Generators (GPHS-RTGs). The Galileo spacecraft (Fig. 2) was launched in 1989 aboard Space Shuttle Atlantis to study Jupiter and the Jovian system from orbit. It was powered by two GPHS-RTGs. The spacecraft arrived at Jupiter in December 1995. A probe was delivered that descended into the giant planet's atmosphere. The orbiter completed many flybys of Jupiter's major moons, reaping a variety of science discoveries. After 34 orbits of Jupiter, the mission ended on September 21, 2003, when the spacecraft was plunged into Jupiter's atmosphere in a planned maneuver for final measurements of the planet's clouds and winds.

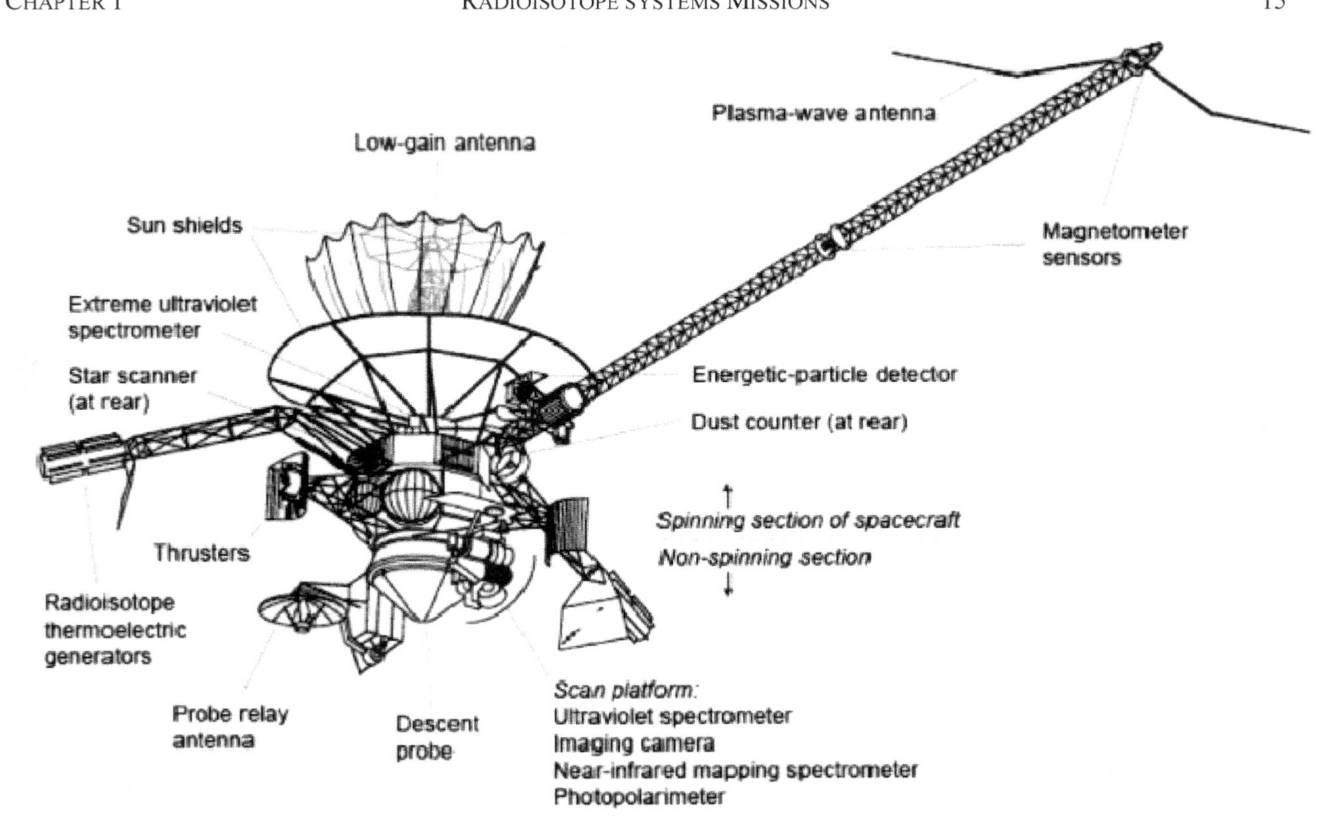

Fig. 2. Galileo spacecraft. *Source: NASA Facts*

The Ulysses spacecraft, launched in 1990 on a European-built spacecraft, used a single GPHS-RTG in its mission to study the heliosphere, the bubble in space blown out by the solar wind. This involves an out-of-ecliptic journey around the Sun. It orbits the Sun once every 6.2 years, making it perfect for studying the 11-year solar activity cycle. The prime mission concluded in 1995; however, Ulysses continued operating until 30 June 2009. It completed nearly three orbits of the Sun during the spacecraft's 18 y of operation.

The Cassini spacecraft was launched in 1997 with three GPHS-RTGs onboard. This provided some 855 watts-electric power--the highest radioisotope powered spacecraft to-date. The spacecraft is the first to explore the Saturn system of rings and moons from orbit. Cassini entered Saturn orbit on June 30, 2004, and drove the Huygens Probe into Titan's thick atmosphere in January 2005. The spacecraft continues to orbit Saturn, sending back new data on Saturn, its moons and rings. In Dec. 2007, it discovered flowing liquid on Titan and continues to map Titan features. The prime mission ended in July 2008, however the mission has been extended.

The latest spacecraft to be launched with a GPHS-RTG is New Horizons. It was launched in January 2006 and is scheduled to reach Pluto and its moons, Charon, Nix and Hydra, in July 2015. As part of an extended mission, the spacecraft will head deeper into the Kuiper Belt to study one or more of the icy mini-worlds in that vast region beyond Neptune's orbit.

The next scheduled use of radioisotope systems is the Mars Science Laboratory. It is scheduled for launch in 2011 and will be a smart rover to assess whether Mars ever was, or is still today, an environment able to support microbial life. It will use a new RTG, called a Multi-Mission Radioisotope Thermoelectric Generator (MMRTG) (Fig. 3), that is designed to operate on planetary bodies with atmospheres such as Mars, as well as in the vacuum of space. The MMRTG contains a total of 4.8 kg plutonium dioxide that initially provides approximately 2,000 watts of thermal power and 120 watts of electrical power. The MMRTG generator is about 64 cm in diameter by 66 cm long, weighs about 43 kg, and is designed for over a 14-year lifetime.

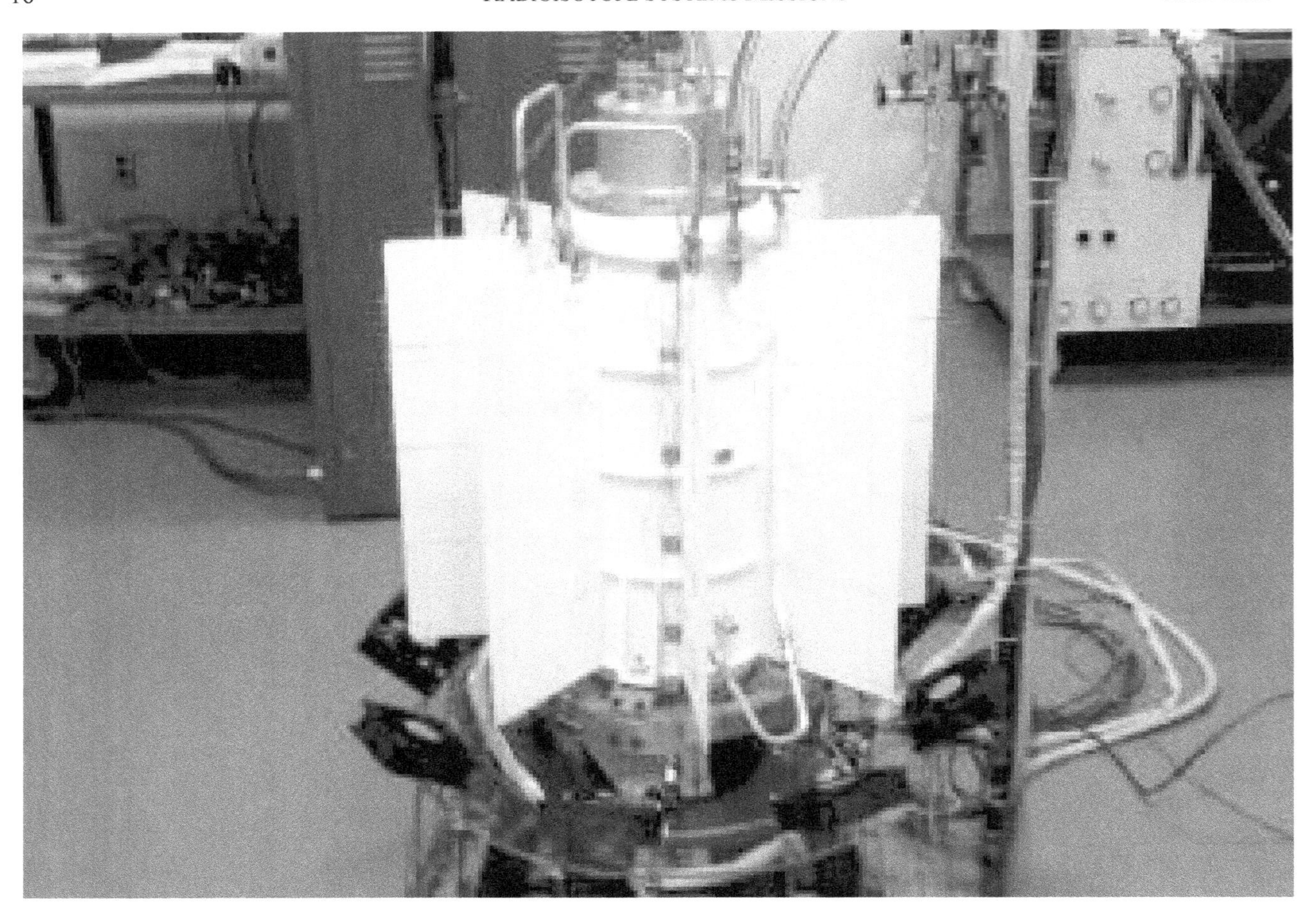

Fig. 3. Multi-Mission Radioisotope Thermoelectric Generator (MMTRG) Engineering Unit. ***Courtesy U.S. Department of Energy***

Fig. 4 illustrates the steady increase in RTG power levels. In 1961, SNAP-3B (the first RTG used in space) had a mass of 2.1 kg and delivered 2.7 watts of electric power to the Transit navigation satellite. The Voyager 1 and 2 missions to Jupiter, Saturn, and beyond on the other hand, had power supplies that delivered 150 watts-electric per generator with a mass of only 37.6 kg. The largest powered spacecraft is the Cassini mission to Saturn. It's three GPHS-RTG power system provided 855 watts-electric and weighed 167.7 kg overall.

Although other radioisotopes have been considered as a heat source, all RTGs flown by the U.S. to date have used the isotope plutonium-238. All these missions have also employed thermoelectric (TE) power conversion techniques with telluride materials in the Nimbus, Pioneer, Viking, Transit, and Apollo RTGs; and silicon germanium (SiGe) converters in the Lincoln Experimental Satellites (LES), Voyager, Galileo, Ulysses, Cassini and New Horizons generators. Fig. 5 shows 16 years of data of the power history of RTGs using silicon-germanium thermoelectric converters. This includes the MHW-RTGs on LES-8/9 and Voyagers 1 and 2 and GPHS-RTGs on Galileo and Ulysses. It illustrates the long term predictable performance and reliability of RTG units.

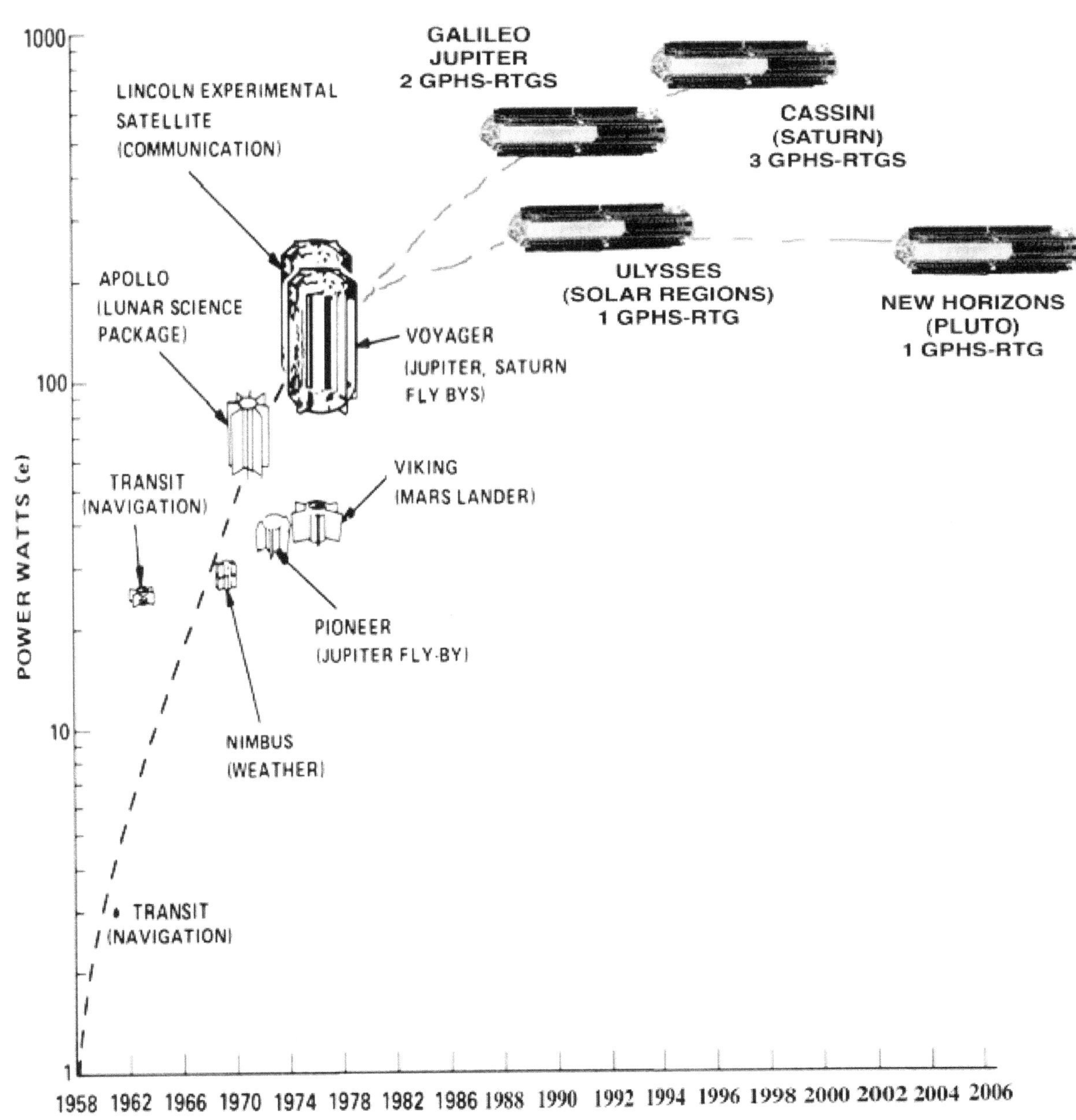

Figure 4. Progress in RTG development. ***Courtesy of U.S. Department of Energy with data added beyond 1986***

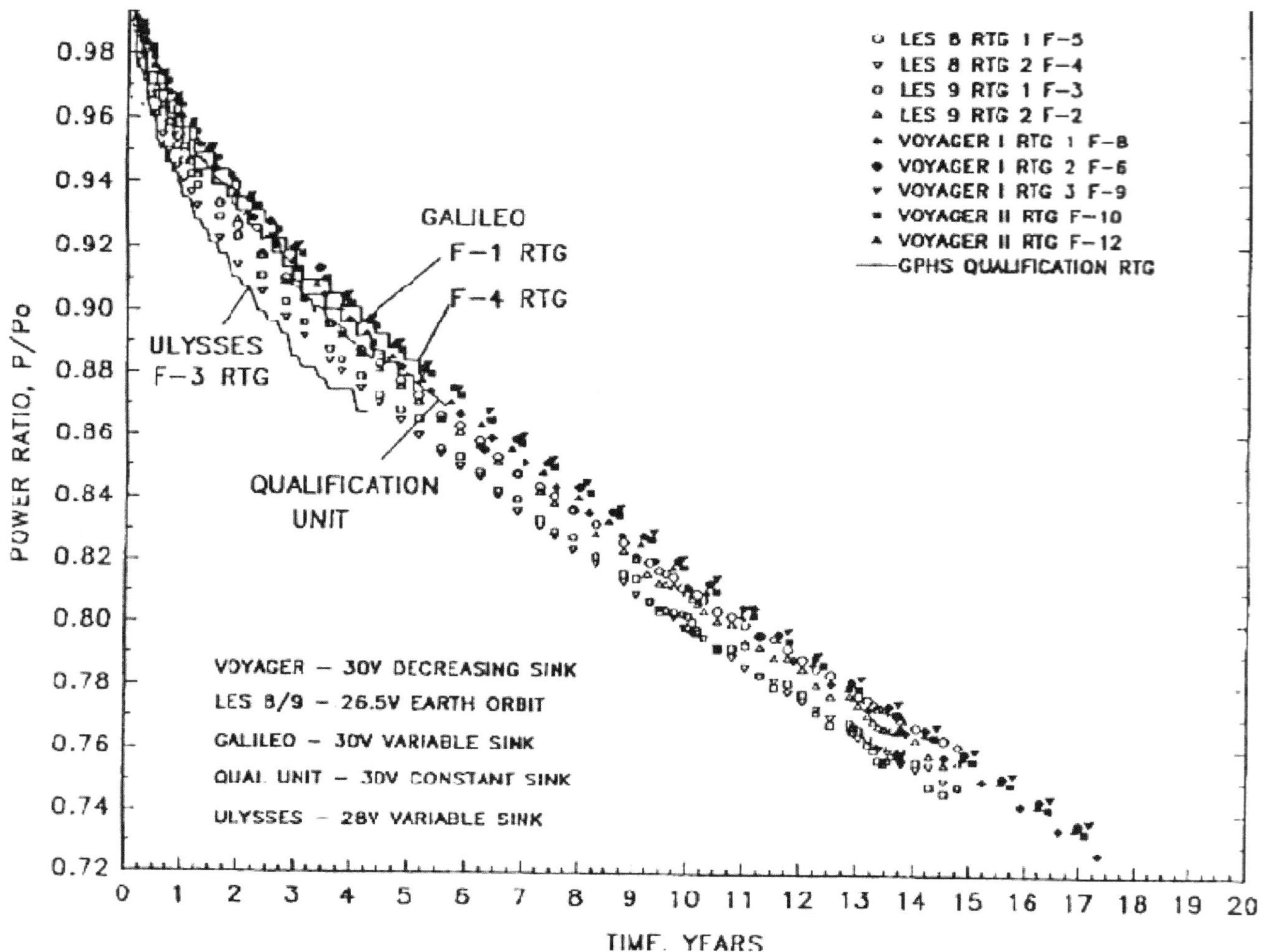

Fig. 5. Power histories to 1994 of all silicon-germanium RTGs (MHW-RTGs on LES-8/9 and Voyagers 1 and 2) and GPHS-RTGs on Galileo and Ulysses. For comparison purposes, the data are presented in terms of the ratio of power at each time to the initial power (P_0). Initial powers for the MHW-RTGs ranged from an average of about 153.7 W_e / RTG on LES-8 to an average of about 159.2 W_e / RTG on Voyager 2. Initial powers for the GPHS-RTGs whose launches were delayed over three years, were about 289 W_e / RTG for Galileo and Ulysses. ***Source: Jet Propulsion Laboratory***

More recent trends in radioisotope power systems and projected developments are shown in Fig. 6. Technology Readiness Levels of 5-6 indicate that the technology is ready for infusion into mission applications. Current technology development is focused on development of a high efficiency (about 28%) Advance Stirling Radioisotope Generator (ASRG) and materials for Advanced Radioisotope Thermoelectric Generators (ARTG). The advanced thermoelectric materials could boost efficiencies from 6.8% to 8 - 10%.[5]

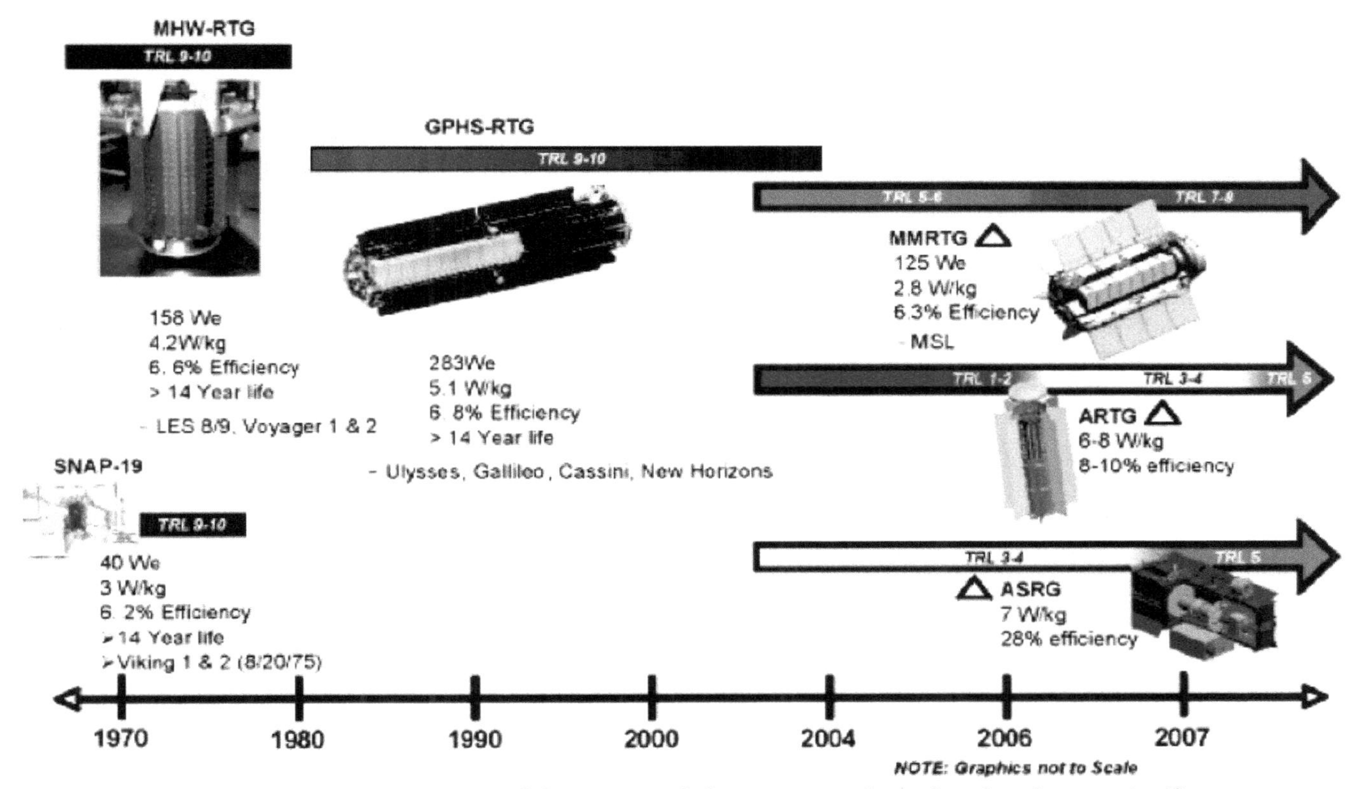

Fig. 6. Radioisotope Power systems of the past and those presently under development. (Source: Alan and Lavery)

Technology Readiness Levels are defined by NASA as follows:[6]

TRL 1 Basic principles observed and reported
TRL 2 Technology concept and/or application formulated
TRL 3 Analytical and experimental critical function and/or characteristic proof-of-concept
TRL 4 Component and/or breadboard validation in laboratory environment
TRL 5 Component and/or breadboard validation in relevant environment
TRL 6 System/subsystem model or prototype demonstration in a relevant environment (ground or space)
TRL 7 System prototype demonstration in a space environment
TRL 8 Actual system completed and "flight qualified" through test and demonstration (ground or space)
TRL 9 Actual system "flight proven" through successful mission operations

In addition to the RTGs, radioisotope heater units have been used to provide heat for electronics and environmental control. For instance, the Mars Pathfinder spacecraft and rover launched in December 1996 and landed on Mars on July 4, 1997, the rover incorporated three Pu-238 oxide heater units. Without the heaters, the Rover would probably not have remained operational after the first Marian night.

In all, the U.S has used a total of 45 RTGs and more than 240 heater units on 26 missions since 1961. As result of the reliability and continuous operation of the RTGs well beyond their design lifetimes, many of the missions have been extended.

Russian Radioisotope Applications[7]

Russia has developed RTGs using Polonium-210; two are still in orbit on Cosmos navigation satellites. COSMOS 84 was launched on 3 September 1965 and COSMOS 90 launched on 18 September 1965. However, since then the Russian have concentrated on fission reactors for space power systems.

As well as RTGs, Radioactive Heater Units (RHUs) are used on satellites and spacecraft to keep instruments sufficiently warm to function efficiently. Their output is only about one watt and they mostly use Pu-238--typically about 2.7g of it. Dimensions are about 3 cm long and 2.5 cm diameter, weighing 40 grams. Two are in shutdown Russian Lunar Rovers on the moon, Luna 17 launch on 10 November 1970 and Luna 21 launched on 8 January 1973.

Chapter 2

Mission Derived Design Requirements And Plutonium-238 Production

The unique challenges of designing space radioisotope power systems are the constraints imposed on the design by the launch vehicles, the high reliability required as a result of not being able to perform any maintenance on the systems once launched, the long lifetimes measured in tens of years required of many of the missions and the special requirements of nuclear safety. Also, one must consider any radiological interaction between the radioisotopes and other components of the spacecraft.

Radioisotope Fuels[1]

The key factors involved in selecting the radioisotope fuel for use in radioisotope power generators are: long half-life compared to the operational mission lifetime; low radiation emissions; a high power density and high specific power (w / kg); and a stable fuel form with a high melting point suitable for the application. Cost of fuel production and cost relative to benefits must be reasonable. Safety considerations in the production and handling as well as potential launch accidents are other important factors. An additional unique requirements has to do with the availability of plutonium-238 fuel for the heat source.

The fuel half-life of a desirable radioisotope to be used in a space system is a function of the mission lifetime. The half-life of the radioisotope fuel should be at least as long, or longer than the mission lifetime, to reduce the power output variation during the mission and to provide a contingency for mission scheduling flexibility. The size and weight of the heat source are directly proportional to the half-life of the fuel. If the half-life of the fuel is too long, the radioactive decay rate is slower and the amount of heat produced per unit time is low. This would result in a heavier fuel source then needed. If the half-life is too short, the isotope decays faster then desirable and therefore requires a larger fuel loading to meet end-of-life conditions.

The radiation environment as a result of the decay of the isotope on the surrounding environment is a source of concern to both personnel and other components of the spacecraft. Alpha particles are most easily shielded, but still can produce energetic neutrons from alpha-neutron reactions with light elements in the fuel. Beta particles produce bremsstrahlung (or x-rays) when they are being slowed down. Spontaneous fission releasing energetic fission neutrons and gammas can results from some radioisotope decay chains.

Dealing with penetrating radiation (gamma, x-rays, neutrons) needs to be avoided. Weigh, being critical in space missions, do not permit the use of heavy external shields. The small size of the fuel pellets and the thin encapsulation materials used in lightweight heat sources for space systems offer very little self-shielding. Personnel must be protection from possible radiation hazards during production, fabrication, testing, and handling before launch. On missions, like the Apollo program, the astronauts were protected while handling radioisotope generators on the Moon. If there is an accident, the general populace must be protected from exposure to radiation levels above those deemed safe. The instruments flown on some scientific missions may include very sensitive particle and photon detectors whose performance must not be degraded by the selected radioisotope heat source.

The power conversion system also influences the selection of a radioisotope fuel. Waste heat must be rejected to space by radiation. This is proportional to the fourth power of the temperature and therefore a high temperature is desirable to reduce the size and mass of the radiator. The fuel must have a stable form (whether compound, alloy, or matrix) compatible with the temperatures in the power conversion system. It must be chemically compatible with its containment material (usually metallic cladding) over the operating life of the heat source. Also, it must be able to with stand high temperatures that could result in postulated accidents, such as fires or reentry heating in the Earth's atmosphere. The fuel form should have a low solubility rate in the human body and in the natural environment.

Daughter products must not adversely affect the integrity of the fuel form nor should the decay process degrade the fuel properties. Helium gas buildup from alpha particle decay must not create inhalable fines within the fuel or fuel cavity which could be released during an accident. The decay process must not destroy the chemical bonds with the fuel form.

Power density (watts / cubic centimeter) and specific power (watts / gram) are two metrics used in establishing the suitability of a radioisotope as a heat source. These are directly proportional to the energy absorbed per disintegration and inversely proportional to the half-life. For comparable power levels, higher power density leads to smaller heat sources and higher specific power leads to lighter heat sources. For radioisotope fuels with comparable half-lives, an alpha emitting heat source will be smaller and lighter than a beta emitting heat source. To manage the helium release from alpha emitting fuels, thin-walled, vented capsules with minimal void volumes or sufficient void volume provided to contain the maximum pressure buildup in the fuel over the life of the heat source are needed.

Daughter products build up as the fuel ages. These can be removed by chemically separation; however, it is best to chemically process the fuel as late as possible before it is to used in the heat source. A more difficult problem is that most isotopes contain some fraction of other isotopes of the desired element that cannot be chemically separated. These dilute the specific power and specific density of the heat source.

Power Conversion Systems

The radioisotope heat source delivers its energy as heat. In some cases, this is used directly to control the environment in the spacecraft or on a planetary surface. However, most of the time, the energy needs to be converted to electricity. Various energy conversion methods exist. Factors such as conversion efficiency, weight, size, operating temperature, reliability, ruggedness to withstand shock and vibration loads, survivability in hostile particle and radiation environments, scalability in power levels, flexibility in integration with various types of spacecraft and launch vehicles, and versatility to operate in the vacuum of deep space or on planetary surfaces are used in the selection of the power conversion system.

Material considerations bracket the selection of the conversion system. Higher temperatures tend to make the power system lighter; however, available materials tend to limit the temperatures that systems can operate reliably over long periods of time.

Reliability is of prime importance. Mission success depends on having sufficient electrical power over the life of the mission. Graceful power degradation over the life of a mission is acceptable as long as it is within predictable limits. Conversion systems with inherent redundancy, such as thermoelectric conversion systems, have been favored to date.

Nuclear Safety [2, 3]

The approach to space nuclear safety is significantly different from terrestrial power plants nuclear safety in the manner of defining postulated accidents. Terrestrial nuclear safety is based on a set of design basis accidents or design events that are defined along with environments. It must be shown that the design can survive these events. Postulated events have evolved over the years with, in a sense, represents a consensus opinion or agreement between the designers and the reviewers. This approach focus resources and makes the safety process manageable and not open-ended. Extreme events, such as core meltdowns, were not part of the required safety analyses.

Space nuclear safety evolved differently then terrestrial nuclear safety. Space safety programs focused on a wide-range of postulated accidents, no manner how unlikely. Probabilistic risk assessment was pioneered by the space nuclear community. A good probabilistic risk assessment spans a wide range of postulated accident scenarios. As such, safety is a prime consideration during the design and operation of radioisotope systems. Comprehensive failure analysis, supported by testing, is used to evaluate postulated failures during ground operations prior to launch, launch operations, failures in orbit, and failures during the mission that might lead to any possible radiological exposure to people. Some safety features incorporated into the system design include: (1) using the surrounding structure to help protect the heat source against explosion fragments and over-pressure during launch pad aborts; and (2) readily disassembling upon atmospheric reentry to free the heat source aeroshells to reenter on their own. In addition, the power system must be able to reject the heat from the radioisotope heat source under all operating and failure conditions without exceeding temperature limits within the heat source.

Every United States nuclear-fueled power supply that is considered for use in space must undergo a rigorous safety review process. This process establishes that the potential risks associated with the nuclear energy source use are commensurate with the anticipated mission benefits. A formalized review process has been developed for evaluating the safety aspects of nuclear system launches. At the center of this process is the Interagency Nuclear Safety Review Panel (INSRP) composed of representatives from the Department of Energy (DOE), the National Aeronautics and Space Administration (NASA), and Department of Defense (DOD). These agencies are responsible for evaluating mission safety for each launch. DOD and NASA personnel are involved because these two government agencies have safety responsibilities and expertise, both as launching organizations and as user organizations of space nuclear power. DOE has statutory responsibility for the safety of space nuclear power systems.

The evaluation process consists of the following elements:

- The lead or sponsoring agency directs the manufacturer of the nuclear power system to write a Preliminary Safety Analysis Report (PSAR) or Updated Safety Analysis Report (USAR) describing all aspects of mission safety.
- The particular version of the Safety Analysis Report is distributed to the members of the INSRP and each member agency conducts its own review and evaluation of the PSAR or USAR.
- A meeting of the INSRP is held with member agencies and their mission hardware contractors (launch vehicle, nuclear fuel, power system, space vehicle, etc.) in attendance. The results of the independent reviews are presented and discussed at this meeting. Action items are generated to resolve any open questions or issues.
- The power system contractor, with input from other agencies/contractors responsible for action items, writes a Final Safety Analysis Report (FSAR) taking into account the FSAR critiques and any appropriate new information.
- Elements 2 and 3 are repeated with the Final Safety Analysis Report (FSAR).
- The INSRP generates a Safety Evaluation Report (SER) that accompanies the request for Presidential approval of the launch.

Fig. 1 shows the generalized sequence of events in this flight safety review process. Because safety features are designed into U.S. nuclear power sources from the very beginning, this safety review process is really an integral part of the overall flight system development and thus should not constrains the overall mission schedule.

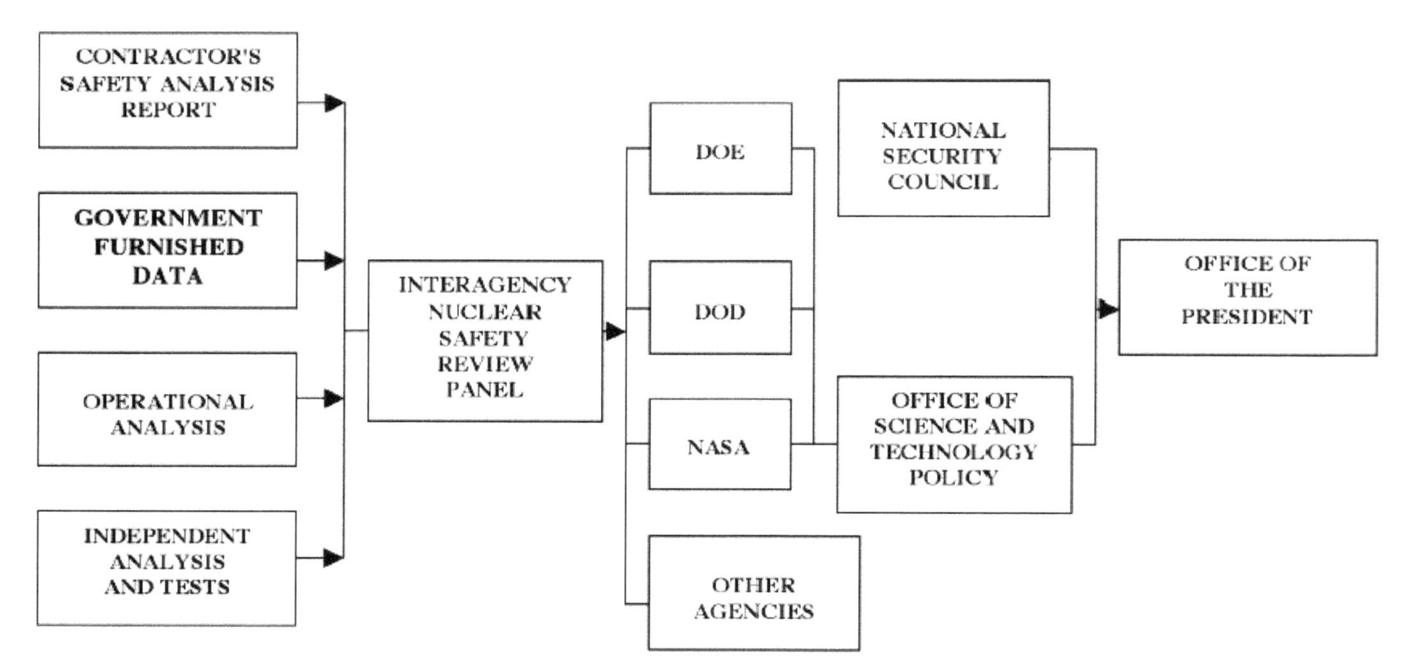

Fig. 1. United States safety review and launch approval process. ***Source: Bennett, 1998***

Detailed safety contingency plans prepared for each mission include potential launch accidents, probabilities, consequences, and risks in various launch phases. This information is documented in the Final Safety Analysis Report. The report typically has the following sections:

- Introduction providing background information on the mission, purpose and scope of the review, and a summary of the Interagency Nuclear Safety Review Process and results of earlier reviews for the mission.
- Mission and system description including the mission profile, launch system, spacecraft and upper stage, and radioisotopic systems.
- Launch site and environments describing the onsite operations, launch complex and exclusion area, and site environment.
- Safety procedures and environment including fire detection and fire suppression, range safety, and contingency plans, procedures and equipment.
- Accidents considered and their probabilities including accident types and failure probabilities.
- Analysis and Conclusions including the INSRP sub-panels, determination of source terms, atmospheric dispersion, health and environmental effects model, accidents analyzed, accidents affecting local Florida area, launch pad explosions, tip over on launch pad, in-flight explosions, rocket failures early in flight, accidents that could affect worldwide health and environment, booster accidents late in flight, orbit accidents, comparative risk assessment, and uncertainties in the analysis.

Sub-panels are used to evaluate the various aspects described above. In the past these have included:

- Launch Abort Sub-panel: to identify and characterize the pre-launch, launch, and ascent accidents, their probabilities, and their associated environments.
- Reentry Sub-panel: to identify and characterize the reentry accidents, their probabilities, and their affects on the nuclear system, including characterization of any postulated fuel releases.
- Power System Sub-panel: to characterize the nuclear system response to pre-launch, launch, ascent, and post-reentry Earth impact accidents, including any postulated fuel releases.

- Meteorology Sub-panel: to characterize the transport of postulated fuel releases within the atmosphere.
- Biomedical and Environmental Effects Sub-panel: to characterize the environmental and health effects of postulated fuel releases, including the overall mission risks.

Risk analysis for a space mission using nuclear power requires:

- Definition of potential mission accidents and probabilities;
- Determination of the types and severity of the resulting accident environments or stresses on the nuclear system;
- Testing and/or analyzing the nuclear system to determine responses to the various accident environments;
- Organization of the information on accidents, probabilities, and system responses into event trees for each mission phase (phases oriented to the potential for human risk);
- Analysis of radiological risk using radionuclide environmental pathways and dose models and world wide data bases; and finally
- Appropriate emergency planning, launch safety preparations, and real time accident analysis and recovery capability.

The logic diagram for the risk assessment is illustrated in Fig. 2.

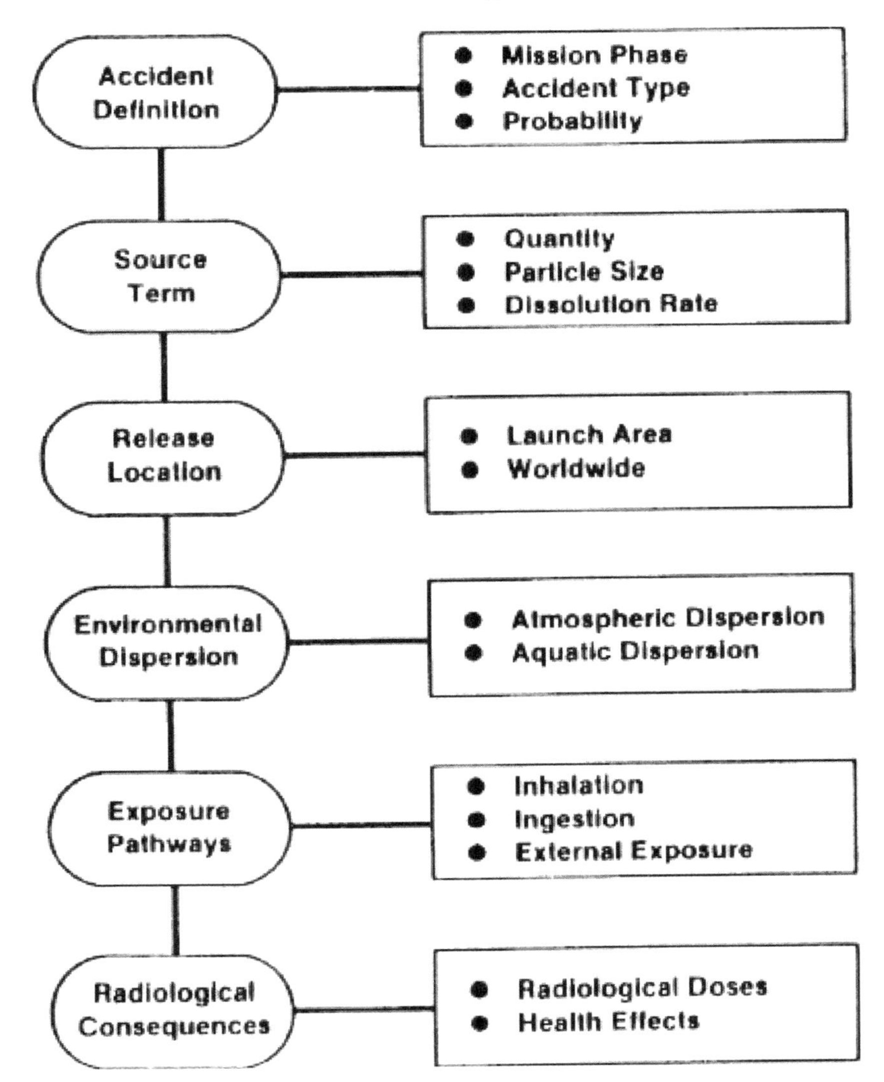

Fig. 2. Logic diagram for analyzing radiological consequences. ***Source: Bennett, 1987***

Safety during ground operations at the launch facility must also be considered in the design process.[4] Radiation emitted by an RTG can affect both personnel and equipment at a launch site. The goal is to keep the radiation dose to all ground workers as low as reasonably achievable (ALARA) and meet Nuclear Regulatory Commission standards. The dose limit for an occupational individual worker radiation exposure allowed is 1.25 rem per calendar quarter. Dose assessments are necessary for all phases of the ground and initial launch operations.

Radiation exposure to the workers who install the RTGs on the spacecraft is unavoidable. However, many other ground operations personnel also have the potential of being exposed. The main effect on equipment occurs during countdown on the launch-pad. Because this is potentially the longest time of exposure, leading to the largest accumulation of radiation dose.

To keep the radiation dose to all workers as low as reasonably achievable (ALARA) without unduly compromising the success of the mission a dose assessment is prepared for each phase of work. This includes time and motion studies of each ground operation and an estimate of the dose rates in those areas where the operations will be performed.

The spacecraft is brought to the launch-pad several weeks before the launch date and mated to the launch vehicle. Installation of the RTGs is delayed to minimize radiation exposure to ground personnel. This has design implications in that the spacecraft design must provide a convenient way to install the heat source late in the installation process. A dose assessment is made for each operation to minimize exposure of launch vehicle personnel. The location of the operations near the RTGs and the time needed to complete operations are estimated. Launch system ground operations personnel may be required to work near the RTGs if there are unplanned contingencies, such as the replacement of components that fail after the RTG has been installed. The estimated dose equivalents for contingency operations ranged from a low of 0.03 person-rem for battery replacement, to a high of 6.4 person-rem for an engine replacement.

A review of equipment on the launch site concerning potential radiation effects must consider both electronics and organic materials. These include such items as cable insulation and Teflon. The exposure time and radiation dose rates to potentially susceptible materials have been found to be low enough that the integrated dose and neutron fluence are below the threshold tolerances for these materials. However, another mechanism that is a potential problem is the generation of bit-flips, called single event upsets, in computer microchips. This can be caused by neutrons coming off the RTG emitted with energies ranging from 1 to 10 MeV. An experimental program to measure these possible single event upsets based on the number of chips and their operation showed a probability of < 0.0005. This indicates the effect is not a problem.

New Horizons Mission

The latest mission to use a radioisotope power generation units is the New Horizons mission to Pluto. This will be used to illustrate some of the factors discussed above in the design process. Fig. 3 shows the major components of the spacecraft.

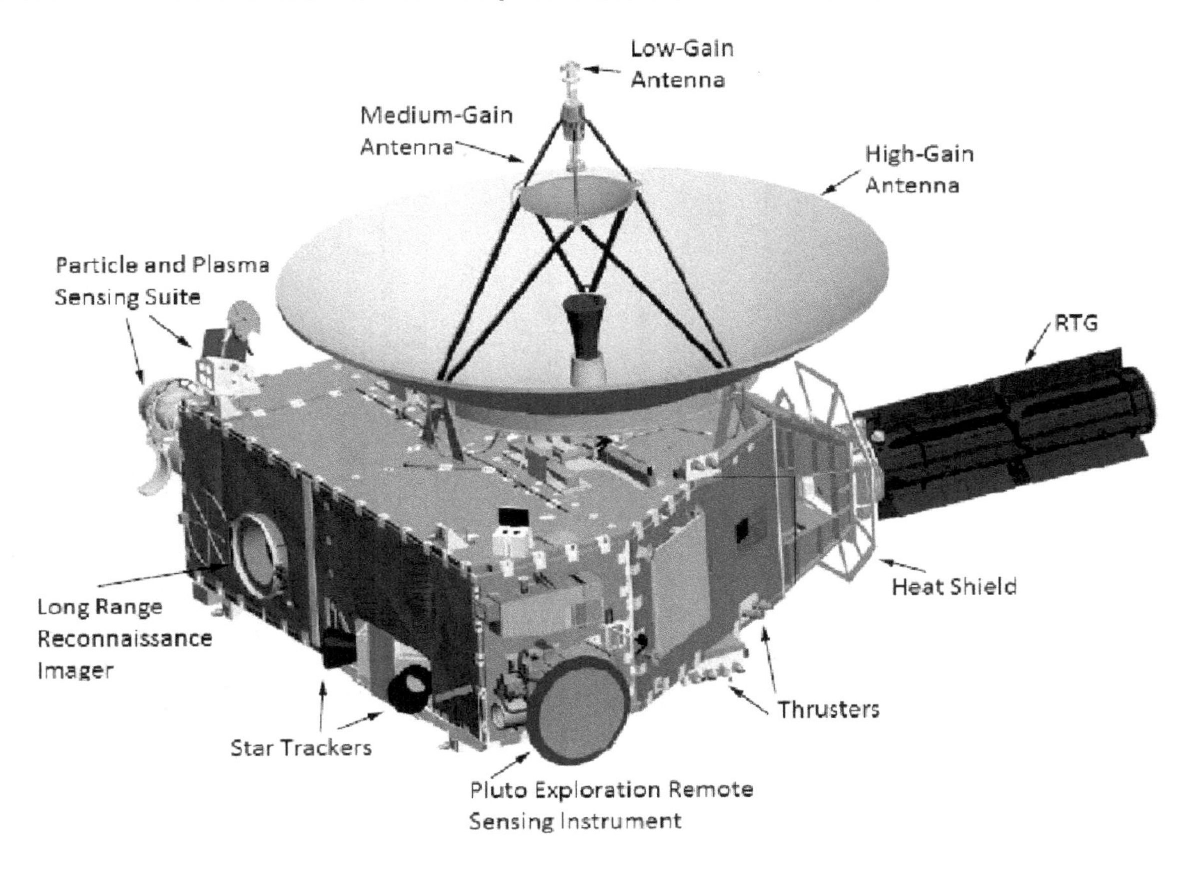

Fig. 3. Major components of the New Horizons spacecraft. ***Source: APL 2003***

The launch systems imposed severe mass and volume constraints on the RTG designs. Current U.S. launch vehicles available for planetary missions are shown in Fig. 4 (the Shuttle is being retired in 2011).

The New Horizons spacecraft instruments are designed to provide information about Pluto's atmosphere content and behavior, the surface of Pluto, and how the solar winds interact with Pluto's atmosphere. Instruments include a visible and infrared imager/spectrometer to provide color, composition and thermal maps; an ultraviolet imaging spectrometer to analyze composition and structure of Pluto's atmosphere and look for atmospheres around Charon and Kuiper Belt Objects; a radio science experiment to measure atmospheric composition and temperature; a long range reconnaissance imager telescopic camera to map Pluto and provide high resolution geologic data; solar wind and plasma spectrometer to measure atmospheric "escape rate" and observes Pluto's interaction with solar wind; energetic particle spectrometer to measure the composition and density of plasma (ions) escaping from Pluto's atmosphere; and a Student Dust Counter to measure the space dust peppering New Horizons during its voyage across the solar system. The instruments established the power levels needed to be supplied by the radioisotope power generation unit.

The launch vehicle was an Atlas V rocket. The instruments, along with the power system and structures, fit into a 479 kg package. With the limitations of the launch capability of the Atlas V, every kilogram used in the power generation system either reduces the number of instruments that can be used for scientific data collection or constrains the size and weight of the selected instruments. Similarly, the goal design of any radioisotope power generation system is constrained by the launch vehicles available for a particular mission.

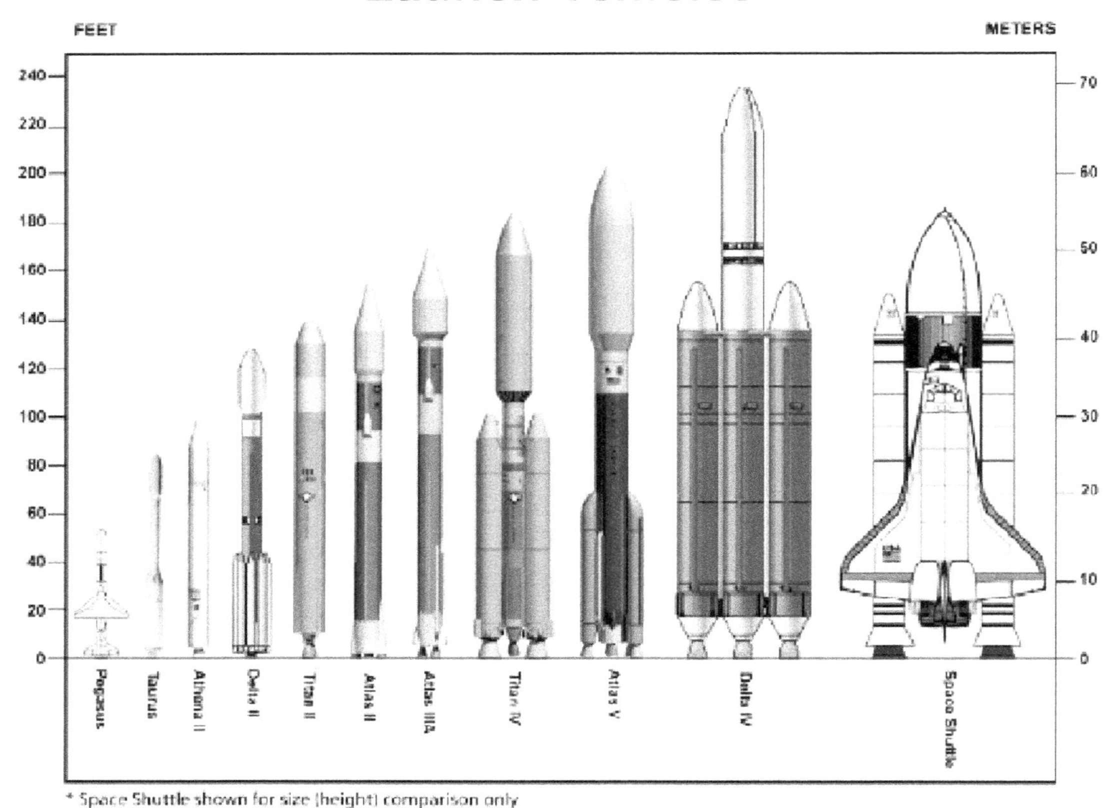

Fig. 4. ***Source: NASA 's Launch Service Program, IS-2005-05-015-KSC (Rev. 03/06)***

The launch vehicle also imposes other constrains on the power generation system design. During launch, the vehicle imposes the most severe forces or loads that the power system will probably experience during its lifetime. The system must also be designed for in-space thrusting loads. Loads experienced during ground transportation can be minimized by the shipping containers and thus should not be limiting design factors.

Lifetimes become a key design criteria. The New Horizons spacecraft, launched in 2006, will not reach Pluto until 2015. After its Pluto-Charon encounter, it will proceed to study the Kuiper belt objects until 2020. Thus, the power system must be designed for an operational lifetime of fourteen years. Since radioisotope heat sources thermal energy declines with time, the electric power output must consider that sufficient power is available at all times in the mission cycle. In addition, there is a certain amount of time before launch for testing and storage that must be considered in the lifetime requirements.

Reliability is a major design driver. Single point failure locations should be avoided. Where this is not practical, sufficient robustness must be shown to mitigate risk of failure.

The spacecraft was launched on 19 January 2006 by an Atlas V 551/Centaur/STAR 48B launch vehicle from Cape Canaveral Air Force Station, Florida. A single GPHS-RTG was used that contained 10.9 kg of plutonium dioxide. Mission operations were divided into the following phases based on the principal launch events.

Considered in the design process was a risk assessment.[5] Phase 0 (Pre-Launch) and Phase 1 (Early Launch): either of these phases could result in ground impact of the RTG and possible release of PuO_2. Before launch, the most likely result of a launch vehicle problem would be a safety hold or termination of the launch. After liftoff, the most significant launch vehicle problems would lead to the automatic or commanded activation of on-board safety systems to destroy the launch vehicle. For both Phases combined, the total probability of an accident is about 1 in 620--this is considered to be in the category of an unlikely event. For individuals within the exposed population, the maximum dose would vary and have a mean value of about 0.3 rem. In the United States, an individual average of about 0.36 rem per year from natural and other sources (such as medical X-rays). Therefore, this is about 80% of normal annual dose exposure. The collective dose received by all individuals within the potentially exposed local and global populations would be about 718 person-rem. (A person-rem considers the population potentially exposed to radiation from an accident scenario and the radiation source term that the population is exposed to.) This results in about 0.4 health effects within the entire group of potentially exposed individuals. Since the lifetime risk of a fatal cancer fatality everyone is expose to is 20%, this added risk in case of an accident is very small.

Phase 2 (Late Launch) accidents would lead to impact of debris in the Atlantic Ocean with no release of PuO_2. The aeroshell modules have been shown to survive water impact at terminal velocity. There are no health consequences.

Phase 3 (Pre-Orbit) is the period before reaching Earth parking orbit and thus could lead to a sub-orbital reentry within minutes. Breakup of the spacecraft during reentry could result in impacts of individual aeroshell modules along the vehicle flight path over the Atlantic Ocean and southern Africa. If the aeroshell modules impact hard surfaces like rocks, small releases of PuO_2 are possible at ground level. The total probability of an accident release is considered unlikely, at about 1 in 1,300. The maximum individual dose is calculated to be about 0.4 rem, or equivalent to 30% of normal background dose. The collective dose is about 3 person-rem, which could result in 0.002 health effects within the exposed population.

Phase 4 (Orbit) include postulated accidents occurring after attaining parking orbit and resulting in reentry following orbital decay. The decay orbit could occur in minutes to years and affect Earth surfaces between approximately 28° North Latitude and 28° South Latitude. Post-reentry impact releases of aeroshell modules would be similar to Phase 3, except more aeroshell modules could impact hard surfaces. This Phase has a computed unlikely release of about 1 in 1,300. The maximum individual dose is about 0.4 rem, or the equivalent of about 110% of normal annual background dose. The collective dose is about 34 person-rem, with the result of about 0.02 health effects within the exposed population.

Phase 5 (Escape) would not result in a release of PuO_2.

The overall total probability of an accident resulting in a release across the entire mission is considered to be in the unlikely category, about 1 in 300. The maximum individual dose is about 0.3 rem and the collective dose about 353 person- rem. The mean health effects (i.e., additional latent cancer fatalities) are estimated to be a small 0.002 to 0.4 within the potentially exposed population.

Plutonium-238 Production and Processing[6, 7]

As mentioned earlier, the availability of Plutonium-238 is a major design factor. Plutonium-238 is not a naturally element. It is produced by irradiation of Neptunium-237 (Np-237) targets in a nuclear reactor. The Np-237 is derived from the production and irradiation of highly enriched uranium fuel. Facilities designed to produce and handle Pu-238 provide shielding for personnel against radiation exposure, confinement of alpha particles, and cooling capability to dissipate a significant heat loading. For Pu-238, criticality control is not an issue because the amount of fuel required to achieve a critical mass, even with a moderator, is very large. Heat loading is the more critical issue.

The United States has not produced Pu-238 since the shut down of the K Reactor at the Savannah River Site in the late 1980's.[8] After exhausting the U.S. supply of Pu-238, additional Pu-238 was procured from Russia from the early 1990's until 2010. This included the purchase of some 30 - 40 kg of Pu-238. All of this was used for NASA missions--it was agreed not to use the purchased Pu-238 for national security applications.

With a generator such as the multi-mission RTG using 3.5 kg of Pu-238 each and taking into account the NASA mission planning model, the estimated future need for Pu-238 is about 5 kg per year. To re-establish the infrastructure to support future U.S. needs has been estimated at several hundreds of millions of dollars.

The nuclear infrastructure required to produce radioisotope power system is composed of: (1) the production of Pu-238; (2) the purification and encapsulation of Pu-238 into a fuel form; and (3) the assembly, testing and delivery of the power source to the users. Fig. 5 illustrates the way the U.S. produced Pu-238 historically.

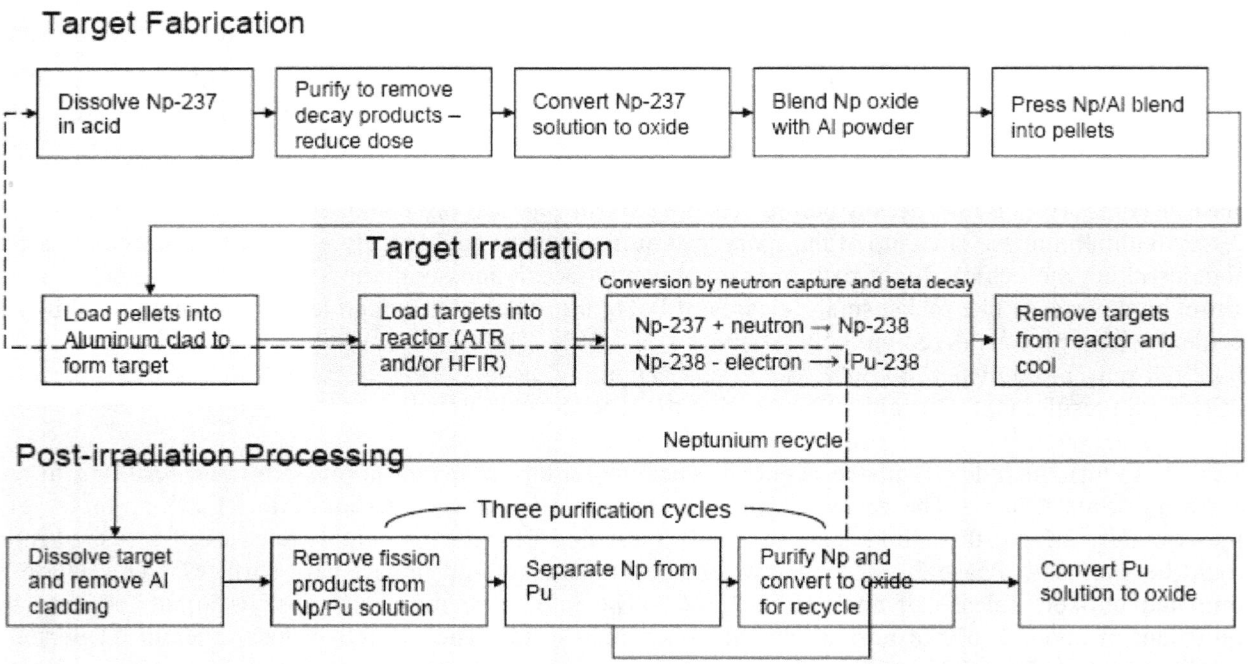

Target Fabrication
Dissolve Np-237 in acid
Purify to remove decay products – reduce dose
Convert Np-237 solution to oxide
Blend Np oxide with Al powder
Press Np/Al blend into pellets
Target Irradiation
Conversion by neutron capture and beta decay
Load pellets into Aluminum clad to form target
Load targets into reactor (ATR and/or HFIR)
Np-237 + neutron → Np-238
Np-238 - electron → Pu-238
Remove targets from reactor and cool
Neptunium recycle
Post-irradiation Processing
Three purification cycles
Dissolve target and remove Al cladding
Remove fission products from Np/Pu solution
Separate Np from Pu
Purify Np and convert to oxide for recycle
Convert Pu solution to oxide

Fig. 5 Historic process flow for Pu-238 production and recovery. Product is plutonium dioxide powder with an isotopic content of Pu-238 greater than 80%. Each production cycle converts 10 - 15% Np-237 to Pu-238 with remainder of Np recycled. ***Courtesy of Dennis Miotla.***

Np-237 targets are irradiated and than chemically processed to extract the Pu-238 isotope. This is purified and converted to an oxide powder (PuO_2) form. The Pu-238 powder is formed into pellets and encapsulated in iridium metal cladding. The encapsulated pellets are assembled into RTG heat sources.

Chemical processing is used to separate the Plutonium-238 from the target assemblies. The assemblies are dissolved in a strong acid solution and the desired product extracted from the resulting liquid solution by various extraction processes. One process often used has a heavy material solution introduced into the top of an extraction column while a light organic solvent material is introduced into the bottom of the column. A combination of gravity moving the heavy material through the solution and mechanical agitation enables chemical interaction to occur. The liquid processing systems is completely closed to provide control of hazardous radioactive liquids. The system is sealed to prevent escape of the dangerous solutions and pressurized to prevent vaporization of organic solvent solutions. Following the extraction process, the fuel is in a liquid nitrate form. This is dried and calcined in a furnace. The water and nitrogen are driven off and the Pu-238 is oxidized.

Fuel fabrication of the oxide powder involves ball milling, sieving, pellet pressing, and sintering. Once the oxide powder is formed into a pellet, the pellet is welded into an iridium cup. The welding process is automated; however, significant handling is required to set up the pellet and hardware and align the welding heads for the automated welding. All of these operations are performed in shielded glove-boxes and provision made to remove the decay heat generated by the radioisotope.

Pu-238 fuel is a strong alpha emitter and therefore process enclosures are designed to provide alpha tight seals at contamination boundaries. Assembly of fueled power systems is performed in shielded hot cell enclosures equipped with master slave manipulators and glove ported windows. Assembly work area use negative pressure relative to the outside of the process enclosure to prevent spread of contaminants into personnel areas.

Production of Pu-238 including the fabrication of neptunium-237 targets, irradiation of the targets in a suitable irradiation facility, and the recovery of Pu-238 from the irradiated targets through chemical processing was formerly performed at DOE's Savannah River Site. However, these reactors are no longer operating. After these reactors were deactivated, Pu-238 was bought from Russia for space missions. This source is about depleted.[9] DOE has decided that when Pu-238 productions is restarted that it will be done at Oak Ridge National Laboratory (ORNL) using the Radiochemical Engineering Development Center for the fabrication of targets and extraction of Pu-238 from the irradiated targets. The Advanced Test Reactor located at the Idaho National Engineering and Environmental Laboratory (INEEL) supplemented by the High Flux Isotope Reactor at ORNL will be used to irradiate the targets. Np-237, the feed material for fabrication of targets for Pu-238 production, has been stored at the Savannah River Site (SRS), but will be stored at INEEL.

The Pu-238 is purified and encapsulated in a metal capsule and welded closed. This work is currently performed at the Los Alamos National Laboratory. The finished capsules are the radioisotope heat source for space systems. Typically, the isotropic composition of the RTG is shown in Table 1

Table 1. Typical isotopic composition of an RTG.

Fuel Component	Weight Percent[(a)]	Half-Life years	Specific Activity curies/gram of fuel component	Total Activity curies
Plutonium	83.63			
Pu-236	0.0000011	2.851	531.3	0.637
Pu-238	69.294	87.75	17.12	129,308
Pu-239	12.230	24.131	0.0620	82.65
Pu-240	1.739	6,569	0.2267	42.97
Pu-241	0.270	14.4	103.0	3,031
Pu-242	0.0955	375,800	0.00393	0.0409
Actinide Impurities	4.518	NA	NA	NA
Oxygen	11.852	NA	NA	NA
Total	100.00	NA	NA	132,465

(a) Based on 10.9 kg (24.0 lbs) of PuO_2 fuel. Source: DOE 2005

NA = Not Applicable

The assembly and test operations, formerly performed at the DOE Mound Site in Miamisburg, Ohio has been moved to INEEL for added security.

In summary, plans exist to resume Plutonium-238 production when funding becomes available. Since certain facilities no longer are operational and to increase security, the production complex has been modified from that used in the past. Other possible production schemes are also being investigated.

Chapter 3

Earlier Generations of Radioisotope Thermoelectric Generators

Basic Building Blocks of Radioisotope Generators

Heat Source[1, 2]

Radioactive decay, the decrease in the number of radionuclides present as a result of their spontaneous nuclear transmutation, is an exponential process governed by the expression

$$dN/dt = -N\lambda \tag{1}$$

where N is the number of atoms of a particular radionuclide
λ is the decay constant (time $^{-1}$)
t is time.

Equation 1 is called the law of radioactive decay. It describes the time rate of change of the radionuclide population. The decay constant (λ) is the probability that a particular radionuclide will disintegrate in unit time. The product $N\lambda$ is called the activity of the radioactive substance and corresponds to the total number of disintegrations per unit time. The negative sign in equation 1 indicates that the activity level decreases with time.

The law of radioactive decay represents a fundamental relationship describing radionuclide behavior. Rutherford and Soddy first postulated the relationship in 1902, as a result of their pioneering work on thorium and its daughter species. Bateman,[3] in 1910, created more general forms of the radioactive decay and growth equations, called the Bateman equations. Plutonium-238 is the radioisotope used in current space radioisotope generators. An expanded decay chain diagram for plutonium-238 is presented in Fig. 1.

The curie (Ci) is used to define the intensity of radioactivity in a sample of material

$$1 \text{ Ci} = 3.7 \times 10^{10} \text{ disintegrations per second (dps).} \tag{2}$$

The curie is approximately the rate of decay or activity of one gram of pure radium and is named after Pierre and Marie Curie. They discovered the element radium in 1898.

The SI (metric), or Systems International d'Unites, unit of activity is called the becquerel (Bq). It is defined as exactly one disintegration per second which, in terms of the curie, can be expressed as

$$1 \text{ Bq} = 2.7027 \times 10^{-11} \text{ Ci.} \tag{3}$$

Both units are commonly found in nuclear engineering and physics.

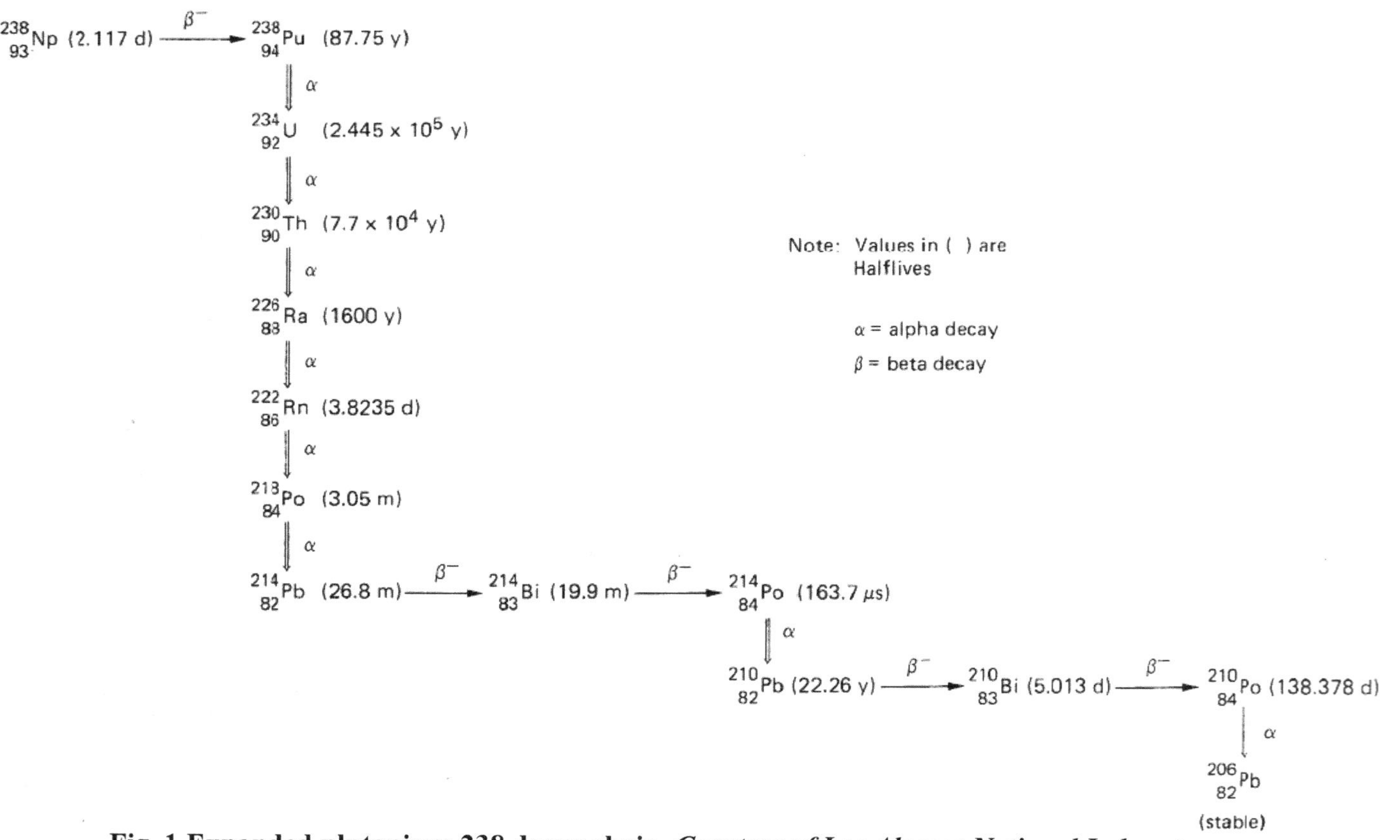

Fig. 1 Expanded plutonium-238 decay chain. ***Courtesy of Los Alamos National Laboratory.***

Equation 1 can be rewritten in integral form as

$$\int \frac{dN}{N} = -\int \lambda\, dt \tag{4}$$

Assuming that the decay constant λ is constant and independent of the age of the radionuclide and that λ is independent of the number of radionuclides present, equation 4 can be integrated to yield

$$\ln\left(\frac{N}{N_0}\right) = -\lambda t. \tag{5}$$

The above important result is more commonly expressed in its exponential form as

$$N = N_0 e^{-\lambda t} \tag{6}$$

where N_o is the number of atoms of a particular radionuclide present at time $t = 0$.

This universal law of radioactive decay applies to all radionuclides. The term, $e^{-\lambda t}$ can be interpreted simply as the probability that a particular nuclide will not decay in time t. The product, $N_o\, e^{-\lambda t}$ represents the total number of radionuclides N that have survived from $t = 0$ to time t.

The half-life $(T_{1/2})$ is a convenient way of representing the rate of decay of a particular radionuclide and is defined as the time in which half the number of radionuclides of a particular isotope disintegrate to another nuclear form. The half-life is related to the decay constant as follows

$$T_{1/2} = (\ln 2)/\lambda \approx 0.693/\lambda. \tag{7}$$

This result is obtained from equation 5 by setting $N = \frac{1}{2} N_0$ and solving for $t = T_{1/2}$. The universal law of radioactive decay is extremely useful for calculating the temporal behavior of typical radioisotope fuels used in space nuclear power systems. The total thermal power level as a function of time, $\dot{Q}(t)$, for a particular (simple) decay chain is given by

$$\dot{Q}(t) = \dot{Q}_0 e^{-\lambda t} \tag{8}$$

where $\dot{Q}_o$ is the initial thermal power level (in watts) at the start of the mission.

A simple decay chain is one in which the decay of the daughter nuclides and beyond may be ignored as a first approximation due to their very long half-lives. For example, the daughter nuclide of plutonium-238 is uranium-234 which has a half-life of 244,500 years. For all practical purposes in calculating the rate of decay of ${}^{238}_{94}Pu$ (the parent nuclide), we can treat its very long lived daughter (${}^{234}_{92}U$) as effectively being stable--at least with respect to its parent. Thus, eq. 6 or 7 is a valid representation of the decay dynamics of a plutonium-238 fueled radioisotope system.

Using these equations, a radioisotope system fueled by plutonium-238 containing one kilogram of plutonium dioxide ($Pu0_2$) with a specific thermal power at the beginning of the mission of 400 watts-thermal for the fuel compound, the total thermal power output of the capsule after ten years of space flight would be 370 watt-thermal. Therefore, after ten years of continuous operation, the radioisotope fuel is still supplying over 92 percent of its original thermal power level to the energy conversion systems.

The plutonium fuel in the heat source is encapsulated for safe handling and to meet safety criteria. This includes a liner, strength member, oxidation resistance cladding, helium vent, and reentry aeroshell.

The liner serves as a decontamination container to prevent radioactive material from contaminating the successive layers of encapsulation during assembly of the heat source. It is a thin metallic capsule next to the fuel form. Tantalum has often been used as the liner material because it is chemically compatible with the fuel over the operating temperatures and lifetimes of interest.

A strength member is incorporated to provide physical containment of the fuel under all operating and accident environments. This includes the ability to withstand high velocity impacts, fragment impacts, explosion overpressure, and fire environments. In addition, since helium is generated as a decay product of the fuel, the strength member may be used to contain the pressure built up in unvented configurations Superalloy materials were used as the strength members for smaller, lower-temperature heat sources designed to burn-up on reentry. Refractory metal strength members are required for the larger, higher-temperature intact reentry heat sources. The refractory metal is protected by a thin noble metal oxidation resistant cladding layer.

To avoid excessive pressure build-up from helium build up, vents are provided in many systems. Design of the vents permits the helium gas molecules to pass without releasing fuel particles or significant amounts of fuel vapor. Several vent designs have been successfully developed and will be discussed under individual generator designs.

An aeroshell member is incorporated when a heat source is designed to meet intact reentry safety criteria. For the aeroshell a good ablative material, such as graphite, is used. The requirement is for low sublimation/ablation rate that can withstand the aerodynamic forces and high thermal stresses experienced during severe reentry heating pulses without coming apart. In addition to preventing metallic fuel components from melting during reentry, the aeroshell must be able to conduct the heat from the fuel to the power conversion subsystem during normal operation.

Nuclear heat source designs have been dominated by aerospace nuclear safety philosophy. Safety philosophy has undergone several revisions since the start of the space nuclear power program. For example, the SNAP-

3B and -9A systems were designed for nuclear fuel burn up and high altitude dispersal in the event of an atmospheric reentry of the nuclear-powered spacecraft; the SNAP-19 system was designed for fuel containment in event of reentry; and the SNAP-27 for containment during reentry and after impact. A key design feature, past and present, is the immobilization of the plutonium-238 fuel during all phases of the mission, both normal and potentially abnormal. This includes possible launch abort, reentry into the Earth's atmosphere, and post-reentry impact.

Early radioisotope generators were based on a design philosophy of "burn-up and disperse". Super alloys (Haynes-25) were selected for use because of the low operating temperatures of the RTG systems. Low temperatures were the result of the choice of lead telluride (PbTe) thermoelectric converters. However, with the advent of higher operating temperature silicon germanium (SiGe) thermoelectric converters and the requirement for intact reentry designs for the RTG systems, the super alloys were replaced by higher melting point noble metals and refractory alloys. Present technology includes iridium alloys that can withstand plutonium oxide ($Pu0_2$) fuel operating temperatures up to 1675 K.

The shape of the nuclear fuel capsules have included right circular cylinders, hemispherical domed cylinders, and spheres. The configurations are based on the overall nuclear heat source design approach, the nuclear fuel form, the converter design, and the properties of the structural materials. To retard adverse material interaction by diffusion and to control emissivities, coatings have been used. This effectively reduced the temperature drops among the various heat source components. Some systems involving plutonium fuel and iridium alloys did not require such protective coatings.

Fuel selection factors that are considered in designing a radioisotope generator include such factors as:

1. The power density (in W / cm^3) of the radioisotope fuel,
2. Its half-life,
3. Fuel availability and its cost,
4. External nuclear radiations associated with the decay of the radioisotope (e.g., neutrons and gamma rays),
5. Contemporary developments in materials science and fuel form technology that directly influence potential radiological and biological hazard of accidentally released fuel, and
6. Potential radiological consequences of the particular fuel should it inadvertently enter the terrestrial biosphere.

Alpha-emitting radioisotopes are generally considered the most attractive nuclear fuels for space applications because of their relatively low shielding requirements and inherently high power densities. The most significant unattractive feature associated with the use of alpha-emitters is helium gas generation. The helium gas has the potential for creating undesirable high pressure conditions within the fuel capsule. Thus, either a strong strength member or a venting system is used. To date, all radioisotope power generators flown in space by the United States have used the alpha emitting isotope plutonium-238. It is readily producible, though costly, and has a long half-life (87.7 years) and a favorable power density.

When using vents, the vents can be either selective or nonselective. Selective vents pass helium, but retain any larger gas molecules and solid particulates from the fuel. Nonselective vents also retain solid particulates from the fuel, but pass helium and other gaseous effluents, including uncondensed fuel, impurity vapors and possibly other fuel decay products such as radon. The SNAP-19 and MHW generators are flight systems that used nonselective vents.

The radioisotope fuel capsule(s) can use many forms of protective layers. As an example, an impact shell can be employed to absorb the kinetic energy of explosion fragments or post-reentry abort impact on Earth. The MHW heat source used a wound-graphite impact shell. Graphite composites have also been investigated to determine whether they can provide significant impact protection in addition to ablation protection during possible launch pad fires, flight aborts, or atmospheric reentry. Graphite ablators have been generally fine-grained ones that offer the best reentry environment ablation resistance, but possess only minimal strength with respect to thermal stresses of impact loads. If the ablator cannot provide sufficient thermal transport during a postulated reentry heating environment, a thermal insulator, such as pyrolytic graphite, can be used to protect the fuel capsule. Such insulating materials surrounding the fuel capsule to meet aerospace nuclear safety requirements creates large, undesirable temperature gradients in the heat source during normal operations. In addition, cladding may be placed around the hot graphite components to protect them from physical damage or from oxidation that might occur during prelaunch handling.

Thermoelectric Power Conversion

Currently, all U.S. radioisotope power systems have used thermoelectric conversion subsystems for converting the heat from the radioisotope heat source to electricity. These devices are considered to be passive power conversion generators in that they have the advantage of having no mechanical moving parts. However, the devices tend to have relatively low power conversion efficiencies--less than 10%.

Thermal energy is directly converter into electricity in a thermoelectric (TE) device based upon the Seebeck effect. Thomas Seebeck (1770-1831), in 1821, observed that an electromotive force (emf) is generated when the junctions of two dissimilar metals are maintained at two different temperatures. Basic to the thermoelectric effect is the fact that a temperature gradient in a conducting material results in heat flow; this results in the diffusion of charge carriers. The flow of charge carriers between the hot and cold regions in turn creates a voltage difference.

Ideal thermoelectric materials have a high Seebeck coefficient, high electrical conductivity, and low thermal conductivity. Low thermal conductivity is necessary to maintain a high thermal gradient at the junction. The semiconductors are connected electrically in series and thermally in parallel.

Practical thermoelectric converters have resulted with the development of special semiconductor materials that combine a high Seebeck coefficient (α), relatively low electrical resistivity (ρ), and low thermal conductance (k). *N*-type and *p*-type semiconductor materials are used in modern TE devices in order to create larger voltage outputs per degree of temperature difference. The basic operating principle of a thermoelectric conversion device is illustrated in Figure 2. An individual TE converter consists of two semiconductor legs that are bonded to two heat transfer surfaces called the hot and cold shoes or junctions. One of the semiconductor materials is a *p*-type material in which the cold shoe region becomes positively charged by the migration of holes under the influence of a thermal gradient. The other semiconductor material is an *n*-type material in which a temperature difference causes electrons to diffuse to the cold shoe. One region of this TE couple is maintained by the bonded hot shoe at the high temperature, while thermal energy is removed from the other end of the couple at the cold shoe. The thermally driven flow of electrons and holes creates a voltage across both cold shoe plates. By connecting an external load across the two cold shoe plates, a current is made to flow through this external circuit. The power flowing through this external circuit is maximized when the load resistance is matched to the internal TE converter resistance. Such TE unicouples can then be connected in an external series-parallel circuit to provide redundant protection against failures that create open circuits.[4, 5]

The output of a TE converter cell depends on the operating temperatures, properties of the *n*-type and *p*-type semiconductor materials, and specific design details. The induced voltage in the cell is equal to the product of the overall Seebeck coefficient (*n* and *p* type material) and the temperature difference. The current flow is equal to the voltage divided by the sum of internal and external resistances. The net output power is equal to the current squared divided by the external load resistance. Again, maximum power occurs when this load resistance is equal to the internal resistance. Thermal energy transferred from the hot shoe to the cold shoe through either semiconductor material leg represents a loss. This conduction process is dependent upon the temperature gradient, the thermal conductivity of the TE materials, and the device design.[6]

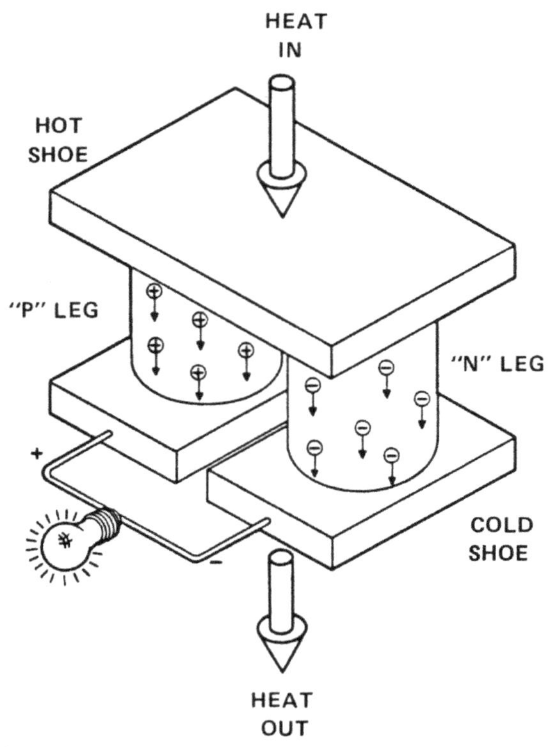

Fig. 2 Operating principle of the thermoelectric converter ***Courtesy R. V. Anderson, et al.***

If the *n* type and *p* type materials have equal thermal conductivities $(k)_{n,p}$ and equal electrical resistivities $(\rho)_{n,p}$, then the figure of merit for our idealized TE converter becomes

$$Z = \alpha^2/k\rho. \tag{9}$$

where α is the material Seebeck coefficient (μV / K).
κ is the thermal conductivity (W / K-cm).
ρ is the electrical resistivity of semiconductor material (mΩ-cm).

The figure-of-merit is usually considered the most significant parameter in the selection of materials for TE power generators. Table 1 and Fig. 3 present figure of merit data for selected TE materials.[7] The larger the value of *Z*, the higher the overall efficiency of the TE converter.

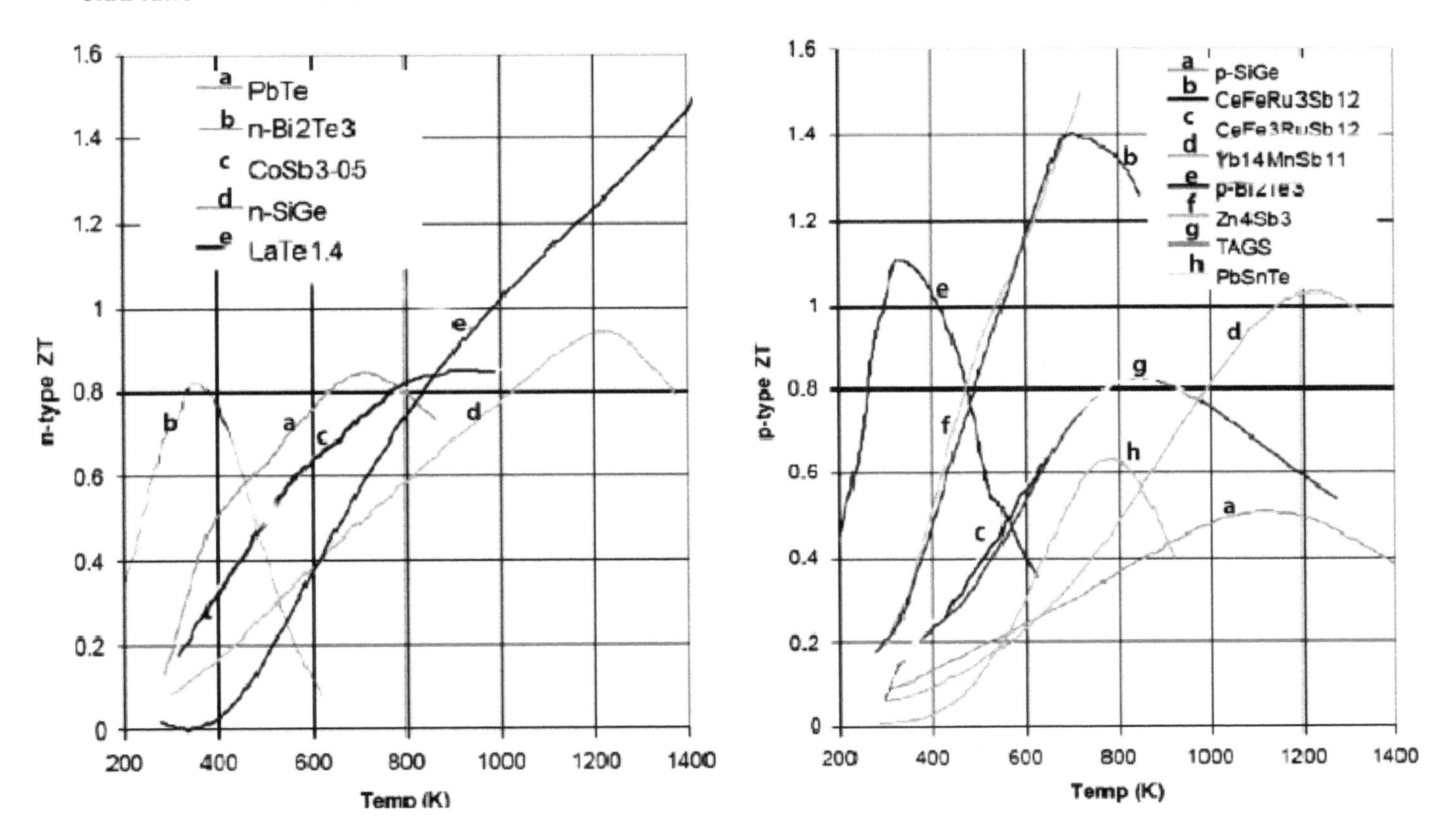

Fig. 3 Figure-of-merit of selected thermoelectric materials. ***From T. Caillat, et al, "Advance Thermoelectric Power Generation Technology Development at JPL," 3rd European conference on Thermoelectric, Nancy, France, Sept. 2005.***

Table 1 Typical calculated figure-of-merit values.

MATERIAL	TEMPERATURE	SEEBECK COEFFICIENT (α)	FIGURE OF MERIT (Z)
METALS	300 K	5 μV/K	3×10^{-6} K^{-1}
SEMICONDUCTORS	300 K	200 μV/K	2×10^{-3} K^{-1}
INSULATORS	300 K	1 mV/K	5×10^{-17} K^{-1}

Table 2 charts the thermoelectric materials used in U.S. Radioisotope Thermal Generators (RTGs).

Table 2. Thermoelectric Materials Used in U.S. RTGs

<u>**TE SEMICONDUCTOR MATERIAL**</u>	<u>**SPACE SYSTEM APPLICATION**</u>
Tellurides	• Transit Navigation Satellites ((SNAP-3A, -9A) • Nimbus III Meteorological Satellite (SNAP-19B) • Pioneer 10, 11 Flybys to Jupiter, Saturn And Beyond (SNAP-19)

	• Viking Mars Lander (SNAP-19) • Apollo Lunar Landings (SNAP-27) • Mars Science Laboratory Rover (pending) (MMRTG)
Silicon Germanium	• Lincoln Experimental Satellites 8/9 (Multihundred Watt Generator-MHW) • Voyager 1, 2 Missions To Jupiter, Saturn And Beyond (MHW) • SNAP-10A U-ZrH Experimental Nuclear Reactor (Earth Orbit) • Galileo Jupiter Orbiter (GPHS-RTG) • Ulysses Planetary/Solar Exploration (GPHS-RTG) • Cassini Orbiter To Saturn • New Horizons On Route To Pluto (GPHS-RTG)

In many configurations the arrays of thermocouples maybe mechanically assembled between the heat source and the heat sink. Springs have been utilized in most telluride TE systems assemblies. These provide good thermal and electrical contacts at the junctions, mechanical forces for withstanding operational or launch payloads and accommodate thermal expansion changes that might occur during thermal cycling of the converter system. Two designs have been created without springs. These are the light- weight, unsealed, bonded panels (launched on the Transit spacecraft) and the hermetically sealed, close-packet tubular module. In SiGe converter systems, such as the MHW unit, the thermocouples are cantilevered from the cold side, and the heat input to silicon molybdenum hot shoes is accomplished by radiative transfer from the heat source.

Thermal insulation is used to reduce heat losses from the ends of the generator and between the thermocouples. Insulation materials depends on the operating temperature, the geometry of the system, and the chemical compatibility of its various components. Good thermal insulation ensures that the thermal energy from the heat source passes through the power conversion thermocouples with losses of less than 10 to 15 percent.

Dynamic power conversion units can also be coupled with radioisotope heat sources to extend power generation range beyond that practically obtainable with TE converter systems and to achieve lower unit costs through higher power conversion efficiencies. However, such dynamic power conversion systems have moving parts that require long-lived bearings for years of maintenance-free operation in space. The most reliable bearings are those that support the moving parts on a film of the working fluid during operation. In addition, dynamic systems have inherent rotational torques and vibrational forces that must be considered in spacecraft integration. Dynamic systems can start up on the ground or in orbit. In the event that the working fluid is lost, an emergency cooling system must be provided to prevent the radioisotope heat source from reaching excessively high temperatures. This emergency cooling system can often be combined in the design of the dynamic power system with the auxiliary cooling systems needed for ground handling and launch operations. Multiple radioisotope heat sources can also be coupled to a common power conversion system or vice versa.

Heat Rejection Subsystem

Waste heat in space systems must be rejected by thermal radiation. Radiator design depends on both operating temperature and the amount of heat to be rejected. The amount of waste heat that can be radiated to space by a given surface area is determined by the Stefan-Boltzmann law and is proportional to the fourth power of the radiating surface temperature (see Fig. 4). This fourth power relationship drives the power generator design to higher temperatures in order to minimize system mass and size. For radioisotope thermoelectric generator power systems, radiator temperatures are usually about 575 K.

In addition, the thermodynamic cycle efficiency is also sensitive to the heat rejection temperature. The lowest possible rejection temperature yields the highest thermal (Carnot) efficiency. Space power plants will usually operate at much lower thermodynamic efficiencies than experienced in terrestrial plants operating on similar cycles. This occurs because a high premium is often placed in space power system design on minimizing radiator area and mass, resulting in the selection of higher heat rejection temperatures. Launch vehicle compatibility is another design constraint placed on the radiator system.

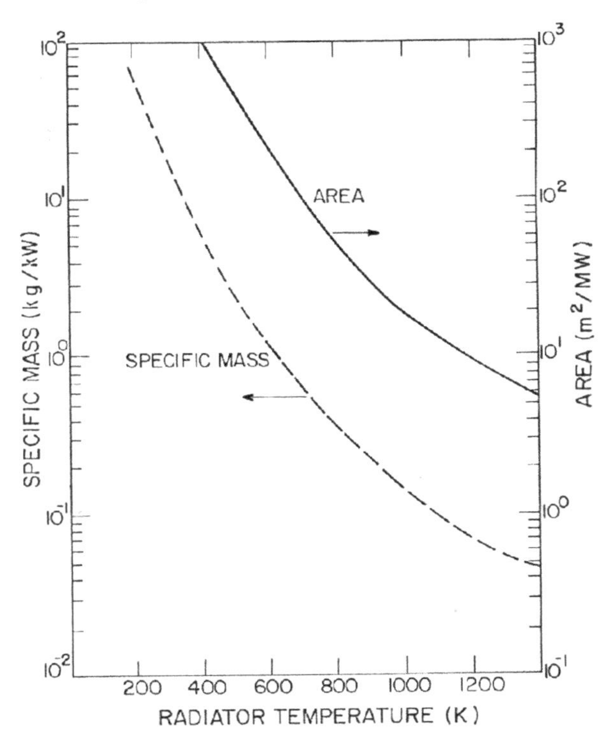

Fig. 4 Radiator area and mass as a function of temperature.

The thermal energy balance for a radiator must consider the following factors: (1) direct solar radiation; (2) Earth-emitted radiation; (3) Earth-reflected solar radiation (2 and 3 are frequently called "earthshine"); and (4) any internally generated heat load from the spacecraft and its nuclear power plant. Earthshine consists of reflected sunlight and thermal radiation from the Earth and its atmosphere. All of these factors represent thermal energy that must ultimately be rejected to space by the radiator. The thermal environment for a spacecraft operating in Earth orbit includes solar radiation (1371 ± 5 W / m^2); Earth radiation (~ 240 W / m^2); and the sink temperature for outer space (~ 0 K) [8, 9]. From the first law of thermodynamics, the overall thermal energy balance equation for the radiator is:[10]

$$\alpha_s F_s G_s + \alpha_r F_r A_p G_s + \alpha_e F_e E_e + P_i / A = \epsilon \sigma T^4 \tag{10}$$

where α_s is the solar absorptivity (0 to 1.0)

F_s is the cosine of the angle between the unit surface normal vector and the direction to the Sun (0 to 1.0)

G_s is the solar radiation incident on a plane normal to the Sun [At one astronomical unit (AU) the solar constant = 1371 ± 5 W / m^2]

α_r is the absorptivity to solar radiation reflected from the Earth (0 to 1.0)

F_r is the view factor for solar radiation reflected by Earth (Typically, F_r = 0.1 for low-Earth-orbit and 0.02 for geosynchronous orbit)

A_p is the Earth's albedo; the fraction of incident solar radiation that is reflected by the Earth and its atmosphere to space (Typical value ~0.3)

α_e is the absorptivity to radiation emitted by the Earth (0 to 1.0)

F_e is the view factor for radiation emitted by the Earth to the radiator surface

E_e is the earthshine radiation (typically ~ 240 W / m^2)

P_i is the internal waste heat load (W)

ε is the emissivity of the radiator surface

σ is the Stefan-Boltzmann constant (5.67 X 10^{-8} W / m^2 - K^4)

A is the area of the radiator (m^2)

T is the radiator operating temperature (K)

Potential radiator materials are a function of the operating temperatures. Fig. 5 shows some candidate materials.

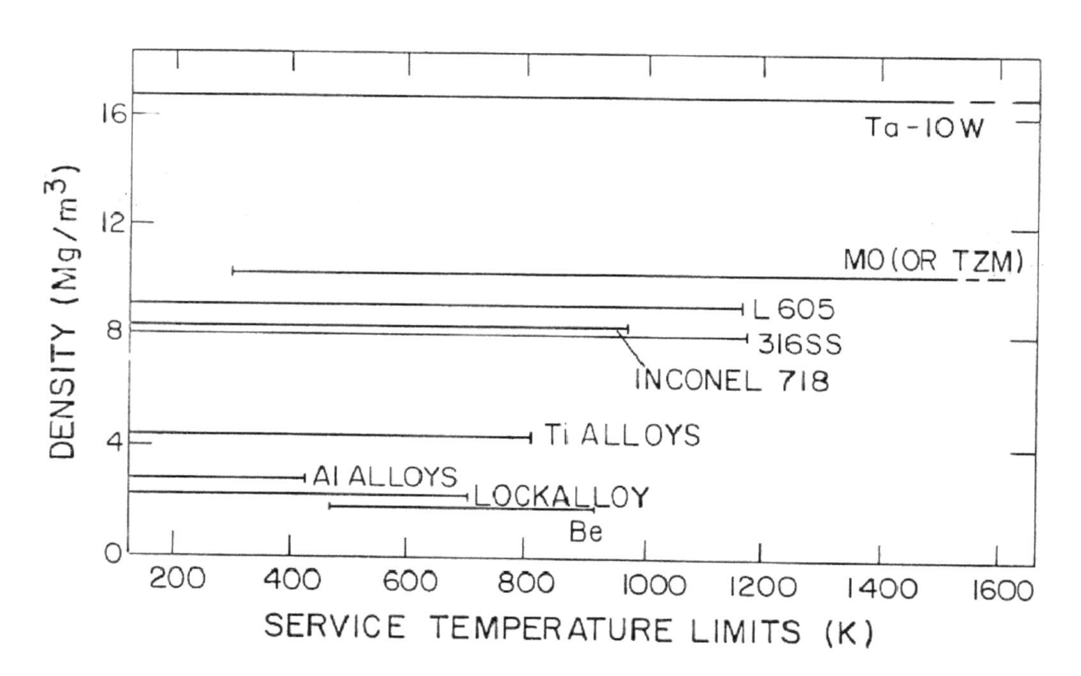

Fig. 5 Potential radiator materials for space power systems.

Heat rejection radiator systems have used static conduction and radiation heat transfer fin designs attached to the generator housing. In order to improve radiator performance, the fins and the generator housing have been covered with coating materials that have a high emissivity-to-absorptivity ratio at normal operating temperatures. Radiator temperatures are a significant factor in the optimization of the system size, power, and mass.

Structure and Thermal Design

A structure is provided for the generator to support the internal parts, attaching the heat rejection system (radiator), and mating with the spacecraft. The structure housing also is a major design consideration for maintaining the inert gas or vacuum operating environment within the generator. Provision in the generator housing is needed for penetrations for the power leads and for any internal monitors.

Because all nuclear power systems first generate heat that is then partially converted into electricity, there are substantial advantages in integrating the power system and the thermal control system in the overall design of the spacecraft. For example, if the power system radiator can be used as the external structure of the spacecraft, mass and volume will be saved. If the heat rejection system is properly designed, external sources of radiant thermal energy (such as the Sun or a nearby planet) will have little effect on the spacecraft's temperature and thermal cycles will be very modest over the entire mission profile. The thermal design of a nuclear-powered spacecraft can also be relatively independent of its orbital altitude or orientation.

Evolution of Radioisotope Generators

The previous chapter discussed mission and safety considerations that govern the design of radioisotope power systems. The evolution of radioisotope generator designs started with a series of power plants called Systems for Nuclear Auxiliary Power (SNAP) and continue to evolute even today. The odd-numbered SNAP power plants use radioisotope fuel: the even-numbered SNAP power plants have nuclear fission reactors as a source of heat. Table 3 presents a detailed summary of the radioisotope generators developed by the united States for space power applications. These radioisotope systems have demonstrated the principles of safe and reliable operation, even though several missions were aborted by launch vehicle failures.

The modern Space Age had its origins in Cold War politics of the 1950s. The successful orbit of Sputnik I by the Soviets in 1957 was viewed as a strategic threat by the United States. At the time, satellite components for in-space use were being developed coincident with the satellites themselves. Reliable, efficient power supplies were crucial, but batteries and solar cells at that time were not optimal for the space environment. Radioisotope power sources were conceived as a power source and led to the development of radioisotope thermoelectric generators.

A combination of the Sputnik I launch, the problem of robust, all-weather naval navigation, and creative thinking led to the Transit satellite program. This became the precursor of the modern Global Positioning System. This initial use of RTGs opened the wider door for the use of such supplies to the exploration of the solar system. Table 4. is a summary of the Transit radioisotope powered RTGs. The TRIAD spacecraft, with its Transit-RTG (RPS), was still operating after 26 years.[11]

Table 1. Radioisotope generators developed for space electric power by the United States

Designation	Application	Electric Power (W)	Weight (kg)	Design Life	Fuel	Status
SNAP-1	Satellite (Air Force)	500		60 days	^{144}Ce	Replaced by longer-lived SNAP-1A in 1959
SNAP-1A	Satellite (Air Force)	125	91	1 year	^{144}Ce	Program canceled in 1959
SNAP-3	Thermoelectric demonstration device (AEC)	2.5	2	90 days	*^{210}Po	Shown to President Eisenhower in 1960
SNAP-3B	Navigation satellite (Navy)(Transit 4A and 4B)	2.7	2	> 1 year	*^{238}Pu	Units *launched* in June and Nov. 1961; June unit operational in 1976; Nov. unit operated to 1971

Table 1. Radioisotope generators developed for space electric power by the United States (con't)

SNAP-9A	Navigation satellite (Navy /Transit 5BN)	25	12	> 1 year	*^{238}Pu	Units launched in Sept. and Dec. 1963 and operated some 20,000 hr before sharp decrease in power; third unit aborted during launch in April 1964
SNAP-11	Surveyor lunar lander (NASA)	25	14	90 days	*^{242}Cm	Surveyor requirement canceled 1965; electrically heated unit delivered to NASA and life tested over 5 yrs; fueled demonstration at ORNL in July 1966
SNAP-13	Thermionic demonstration device (AEC)	12.5	2	90 days	*^{242}Cm	Fueled demonstration at ORNL in Nov. 1965; program completed in 1966
SNAP-17A and 17B	Communication satellite (Air Force)	30	14	> 1 year	^{90}Sr	Design and component test phase completed Nov. 1965
SNAP-19A	Imp satellite (NASA)	20	8	> 1 year	^{238}Pu	Design and component tests completed in 1963; not used on Imp because of radiation interference with payload instrumentation
	Various satellites (AEC)	250		> 1 year	^{90}Sr	Six different design studies completed in 1964
	Surveyor lunar roving vehicle (NASA)	40	10	1 year	^{238}Pu	Design and integration study completed in 1964
	Extended Apollo missions (AEC)	1500		30-90 days	^{210}Po	Design and feasibility studies completed in 1964
SNAP-19B1 2, 3	Meteorological satellite (NASA) (Nimbus III)	30	14	> 1 year	*^{238}Pu	First Nimbus launch aborted in 1968 and fuel recovered from off shore waters; replacement unit launched in April 1969, operating continuously at gradually decreasing power level
SNAP-25	Various satellites (AEC)	75	16	> 1 year	^{238}Pu	Program canceled
SNAP-27	Apollo Lunar Surface Experiment Packages (NASA)	63.5	31 (excludes 11 kg cask)	1 year	*^{238}Pu	First unit placed on lunar surface by Apollo 12 astronaut in Nov. 1969 and continuously powered ALSEP well beyond design life; second unit landed in deep Pacific Ocean on Apollo 13 abort. Additional units used on Apollos 14 to 17
SNAP-29	Various satellite and lunar missions (DOD and NASA)	400-500	180-225	90 days	^{210}Po	Program canceled in 1969 with completion of component tests
	Various missions (DOD and NASA)	250		5 years	^{238}Pu ^{90}Sr	Three design studies completed in 1967; controlled intact re-entry and ground handling included
	Thermionic module development (AE)	100	9		^{244}Cm	Program suspended after completion of preliminary design in 1967 and component tests in 1970.
	Isotope Brayton ground test (AEC and NASA)	3000-15,000	-	> 1 year	^{238}Pu	Preliminary heat-source designs completed in 1966; fuel capsule development 1967-1970; NASA completed 2,500 hr life test on electrically heated system in 1970.

Table 1. Radioisotope generators developed for space electric power by the United States (con't)

Designation		Electric Power (W)	Weight (kg)	Design Life	Fuel	Status
Transit RTG	Navigation satellite (Navy) improved Transit	30	14	5 years	*^{238}Pu	Launched in 1972; still operating.
SNAP-19	Pioneer F and G (Jupiter flyby) (NASA)	30	14	3 years	*^{238}Pu	Launched in 1977; missions ended 3/31/97 and 9/30/95.
SNAP-19	Viking Mars Lander (NASA)	35	14	> 2 years	*^{238}Pu	Launches in Aug. and Sept 1975; operated well beyond design life.
Multi-Hundred Watt RTG	Various missions (DOD and NASA)	100 - 200	-	5 - 10 years	*^{238}Pu	Used in LES 8/9 in 1976 and Voyager 1 and 2 in 1977. Voyagers still operating beyond Saturn, Uranus, Neptune.
Dynamic Radioisotope Power Systems (DIPS)	Various missions (DOD and NASA)	500 - 2,000	215	7 years	^{238}Pu	Program 1975 to 1980 demonstrated 11,000 hr Rankine cycle including 2,000 hr endurance test.
GPHS-RTG	Various missions (DOD and NASA)	290	54	40,000 hr	*^{238}Pu	Used on Galileo (1989), Ulysses (1990), Cassini (1997), and New Horizons (2006) missions. All but Galileo still operating. Galileo mission ended with planned termination.
DIPS Brayton Isotope Power Systems	Various NASA missions	500 -1000	246 kg for 500 W, 319 kg for 1,000 W	12 y	^{238}Pu	1998-2001. Designed Integrated System Test.
Alkali Metal-to-Thermoelectric Converter (AMTEC)	Various NASA missions	120	13.6	14 y with 3 y storage	^{238}Pu	Technology programs in 1990s-2000s. At TRL-3 level. BASE, electrodes and current collectors demonstrated >20 y life.
Advanced Thermoelectric Generator	Various NASA missions			14 y with 3 y storage	^{238}Pu	Technology program initiated in 2002. Demonstrate four-couple modules and > 1 y life by 2009.
Stirling Radioisotope Generator	Various NASA missions	151	22	14 y with 3 y storage	^{238}Pu	Technology program initiated in 2002. Advanced Stirling technology being matured to perform 1 y engineering unit test and >25,000 h life in 2009.
Multi-Mission Radioisotope Thermoelectric Generator	Various NASA missions	120	43	> 14 years	*^{238}Pu	Planned launch on Mars Science Laboratory in 2010.
	* Units fueled					

Table 2. Transit Radioisotope Powered Spacecraft. ***Source: Dassoulas and McNutt***

Spacecraft	Designation	Launch	Mass (kg)	Source	Power (W)	RPS Disposition
4-A	1961 o1	29 Jan 1961	78.8	SNAP-3B7	3	Operated 15 yrs.
4-B	1961 $\alpha\eta$1	15 Nov 1961	89.5	SNAP-3B8	3	Operated 9 yrs.
5BN-1	1963-38B	28 Sep 1963	69.3	SNAP-9A	30	Operated 9 mos.
5BN-2	1963-49B	5 Dec 1963	74.7	SNAP-9A	30	Operated 6 yrs
5BN-3	-	21 Apr 1964	75.1	SNAP-9A	30	Launch failure
TRIAD	1972-69A	2 Sep 1972	93.9	RPS	30	RPS operating

SNAP-3B

Nuclear power systems were first used in space in 1961 with the successful launches of the Transit 4A (June) and Transit 4B (November) navigational satellites. The units, SNAP-3B RTGs, supplemented the spacecraft solar cell arrays and operationally demonstrated the use of nuclear systems for space power applications. The spacecrafts operated in a 1,100 km orbits around the Earth. With this altitude orbital lifetimes are greater than 1,000 years. The aerospace nuclear safety philosophy at that time was to design the generators for burnup and high altitude fuel dispersal. The safety level selected in case of a mission abort that resulted in atmospheric reentry of the spacecraft was to reduce radioactivity levels below those resulting from atmospheric nuclear weapons testing.

The SNAP-3B heat source (see Fig. 6) incorporated plutonium-238 in the form of plutonium metal. When used as a radioisotope heat source, plutonium metal contains approximately the following isotopic mixture: ^{238}Pu (80 wt %), ^{239}Pu (16 wt %), ^{240}Pu (3 wt %), and ^{241}Pu (1 wt %). The amount of fuel was 92.7 grams and the total heat source mass 359 grams. The fuel generated 52.5 watts-thermal, and had an operating temperature of 810 K.[12] Tantalum was used as the liner material. A strength member consisted of a Haynes Alloy No. 25 tapered cylinder (see Fig. 7). The SNAP-3B generator used lead telluride (PbTe) TE converter materials. Overall generator efficiency was 5 percent. Springs were used to maintain thermal contact between the nuclear heat source and the TE converter.

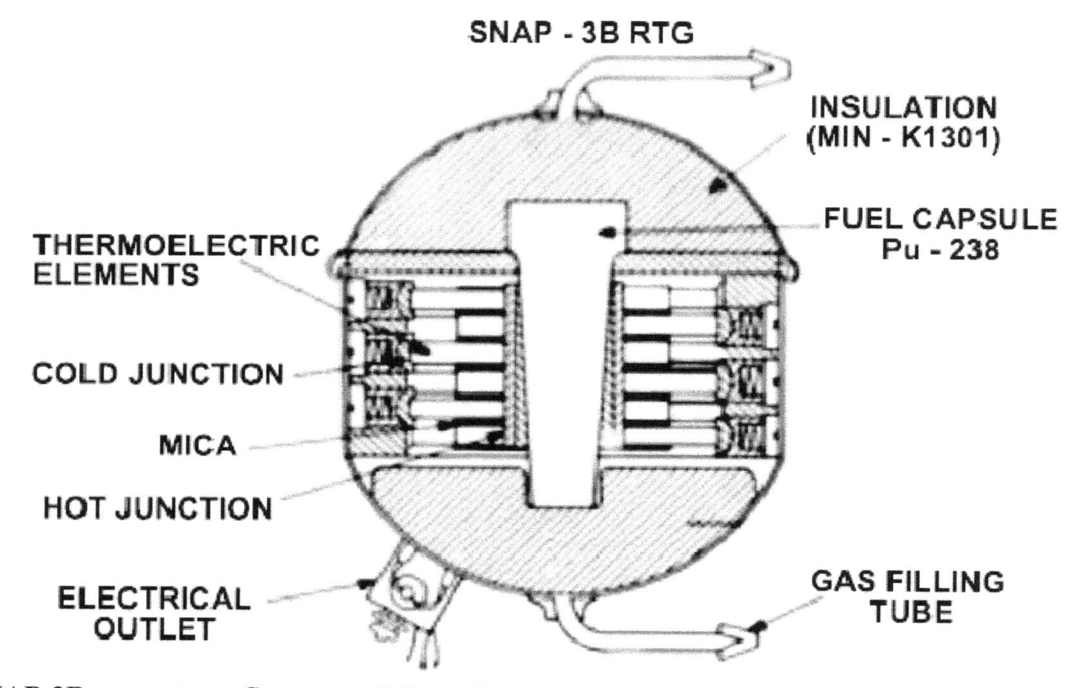

Fig. 6. SNAP-3B generator. ***Courtesy of Mound Laboratory Monsanto Research Corporation***

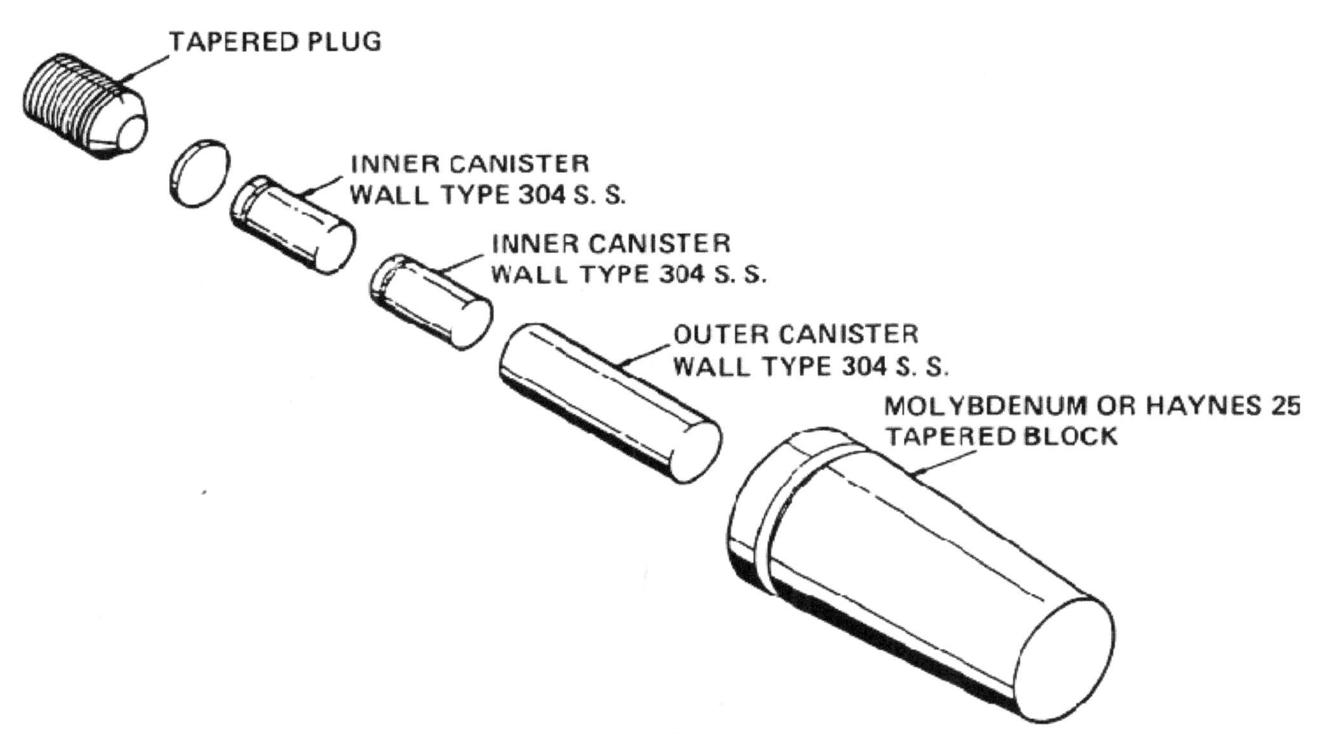

Fig. 7. SNAP-3B fuel capsule. ***Courtesy of Mound Laboratory Monsanto Research Corporation.***

Fig. 8 presents SNAP-3B generator performance data as telemetered to Earth from the Transit 4A and 4B spacecraft. The Transit 4A generator was still operational in 1976, 15 years after its launch, and the last reported signal from Transit 4B was in April 1971.[13]

SNAP-9A

To increase the power output from that provided by the SNAP-3B, the SNAP-9A system was developed. The SNAP-9A, shown in Fig. 9, produced 25 watts-electric. Details of the generator are illustrated in Fig. 10. The generator's nuclear heat source consisted of six plutonium-238 fueled capsules arranged symmetrically about the vertical center line of the generator. The SNAP-9A heat source was 14.6 cm long and 2.5 cm in diameter. An outer capsule was made of Haynes Alloy No. 25 (Haynes 25, for short) and an inner liner of tantalum. Each fuel capsules weighed 458 grams.

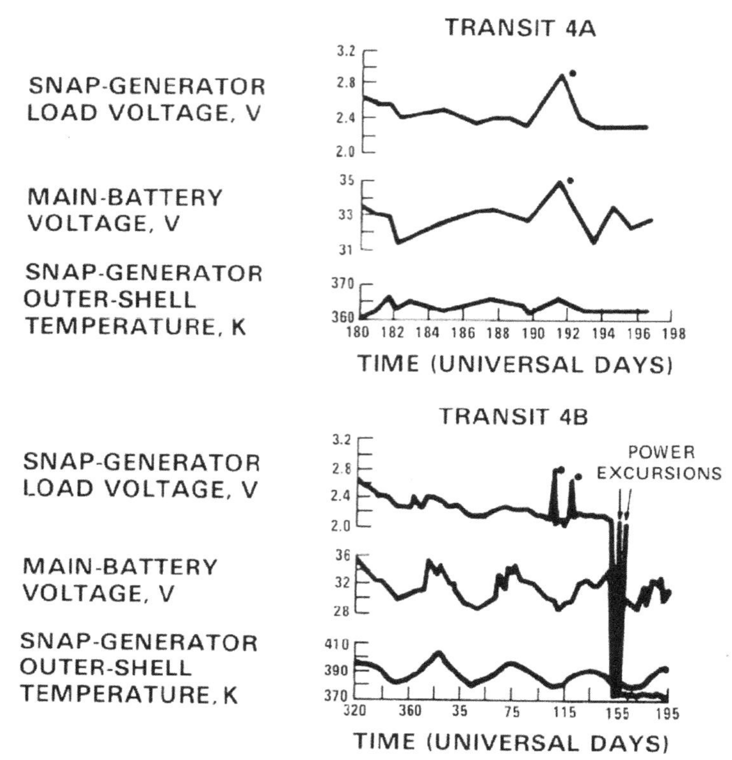

Fig. 8. SNAP-3B generator performance data as telemetered from the Transit 4A and Transit 4B spacecraft. ***From Bennett 1983***

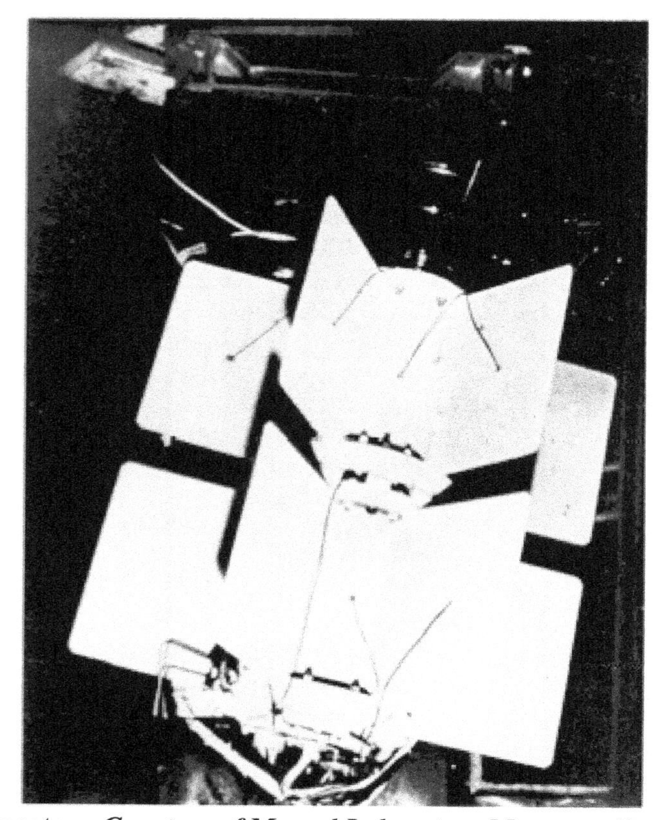

Fig. 9. SNAP-9A generator. ***Courtesy of Mound Laboratory Monsanto Research Corporation.***

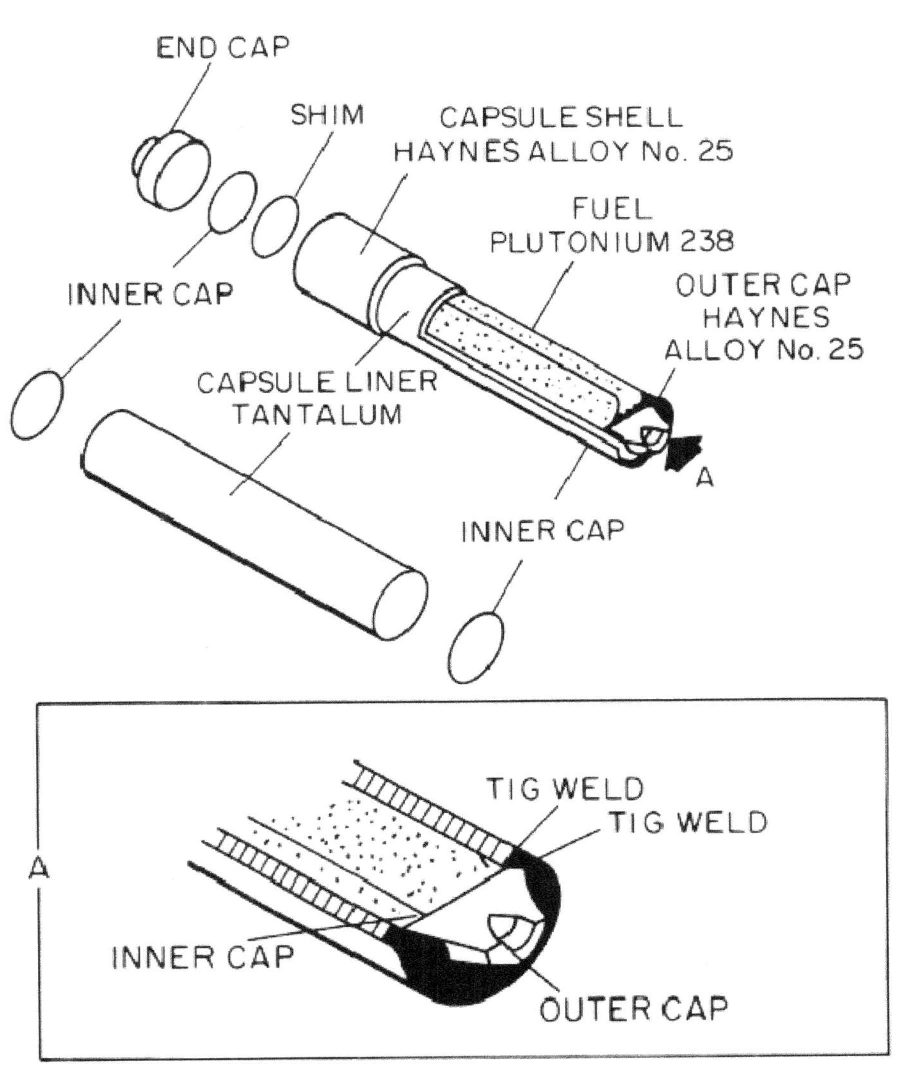

Fig. 10. SNAP-9A fuel capsule. ***Courtesy of Mound Laboratory Monsanto Research Corporation.***

In constructing the SNAP-9A, after fueling, each capsule was placed on its side and was heated, causing the fuel to melt. The melting temperature of the plutonium fuel form was 890 K. Upon solidification, the fuel bonded itself to the capsule liner and established a very good heat transfer environment. Segmented graphite block was used to position the capsules and conducted the heat to the thermoelectric converters. Consistent with aerospace nuclear safety philosophy of the day, in the event that a mission abort caused the spacecraft to reenter the Earth's atmosphere, the fuel capsules were designed for intact impact under launch abort conditions and for high altitude burnup and dispersal. Thus, the segmented fuel block design permitted separation of the capsules for exposure to aerodynamic heating during a reentry abort. The liner was composed of super alloy Haynes 25 with a melting point of about 1650 K. This material accommodated atmospheric burnup, yet resisted oxidation and marine environment corrosion. A tantalum liner was used as a compatibility barrier between the plutonium-238 metal and the Haynes 25 structural member.

Table 3 summarizes the operating parameters for the SNAP-9A system. The peak fuel temperature was 907 K and the temperature drop across the thermoelectric converter was 339 K. Fig. 11 illustrates the SNAP-9A power performance. The top plot in Fig. 11 presents smoothed data from the Transit 5BN-I spacecraft; the bottom plot from the Transit 5BN-2 spacecraft. The gap in the Transit 5BN-2 data corresponds to the period of time when data was not taken.

Table 3. SNAP-9A operating parameters.

Parameter	Beginning of life	End of Life in Space (6 years)
Skin Temperature (K)	430	425
Cold Junction Temperature (K)	450	444
Hot Junction Temperature (K)	789	767
Fuel Centerline Temperature (K)	907	839
Heat Input (watts)	525	500

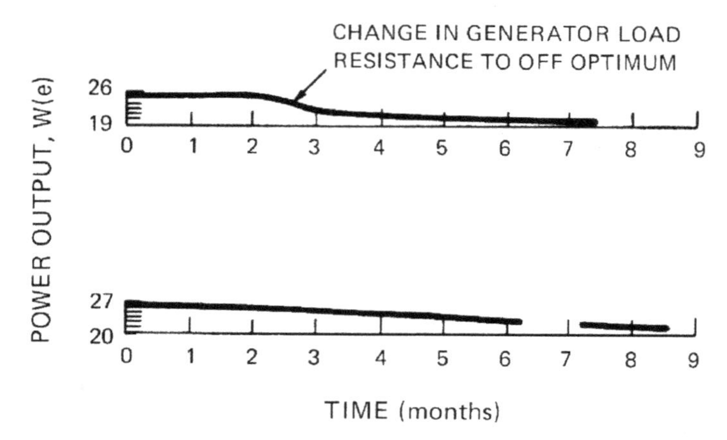

Fig. 11. SNAP-9A power performance data: Transit 5BN-1 (upper plot) and Transit 5BN-2 (lower plot) ***From Bennett 1983.***

SNAP-19

The SNAP-19 systems used a new 645 watts-thermal heat source called the Intact Impact Heat Source. Power conversion consisted of an array of 90 lead telluride (PbTe) and silver antimony germanium telluride (PbTe-TAGS) TE converter materials. A modified safety philosophy was adopted; one to contain the radioisotope fuel under normal operational conditions and to limit the probability of contamination of the biosphere under abnormal launch and mission abort conditions. Safety design requirements included survival upon reentry and impact and containment or immobilization of the fuel.[14, 15 16, 17, 18]

The NASA Nimbus III meteorological satellite, successfully launched in April 1969, was the first satellite to use the SNAP-19 generator. This RTG system operated properly for over twice its design life (approximately 20,000 hours) before experiencing a sharp degradation in performance (see Fig. 12). A decline in performance was attributed to the sublimation of the thermoelectric material and loss of the hot junction bond due to internal cover gas depletion.

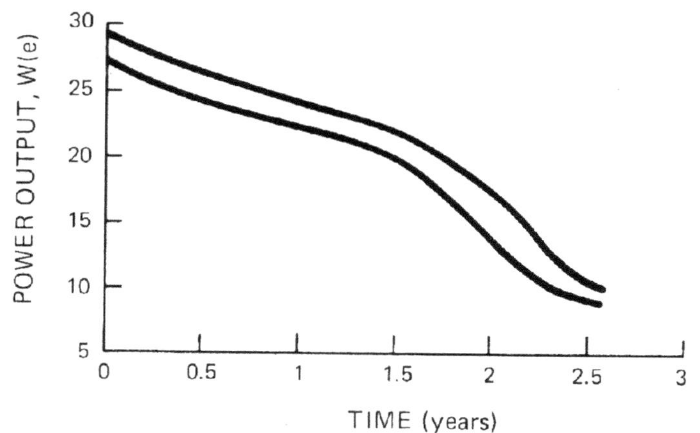

Fig. 12. SNAP-19 power output for the Nimbus III spacecraft. ***From Bennett 1983***

For the Pioneer missions to Jupiter and Saturn and the Viking missions to Mars, the SNAP-19 generator design was modified from the Nimbus III spacecraft configuration to provide up to a six year design life. Among all the deep space travelers launched by the United States, perhaps Pioneers 10 and II are the most appropriately named--for they truly performed pioneering exploration of the outer regions of our Solar System. Pioneer 10 and 11, launched in 1972 and 1973 respectively, were the first spacecraft to fly past Jupiter (in 1973 and 1974). At Jupiter, Pioneer II's trajectory was carefully targeted to swing it (now renamed Pioneer Saturn) toward Saturn for an encounter in September 1979. On 13 June 1983 Pioneer 10 became the first man-made object ever to leave the Solar System. A schematic of the Pioneer 10 and 11 spacecraft is shown in Fig. 13. Table 4 summarizes some of the discoveries made by these automated, nuclear-powered explorers. The SNAP-19 generators continue to operate successfully for over four times their six year design life (see Fig. 14) with Pioneer 10 operating until March 31, 1997 and Pioneer 11 until Sept. 30, 1995.

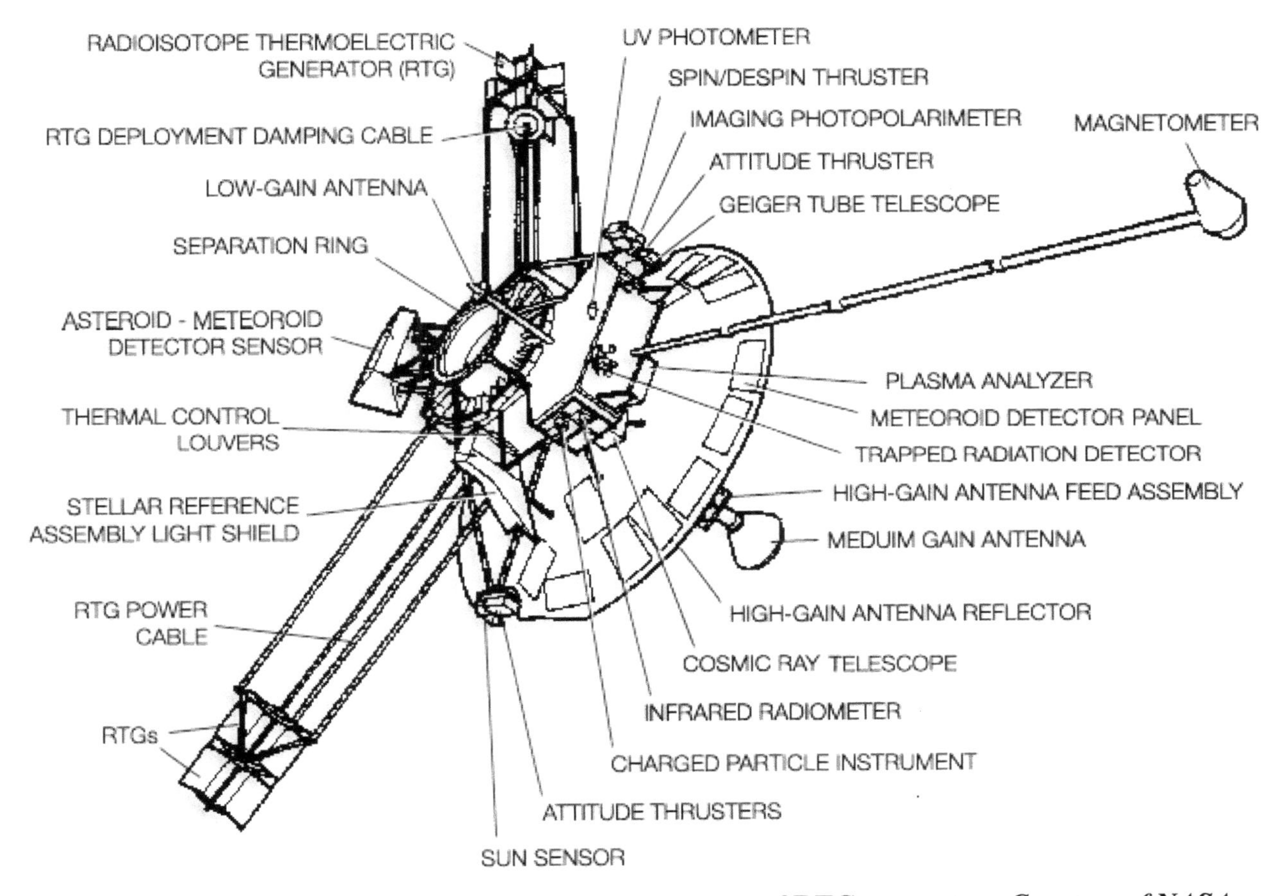

Fig 13, Pioneer 10 and 11 spacecraft showing placement of RTG generator. ***Courtesy of NASA.***

Table 4. Selected scientific highlights of the Pioneer missions to Jupiter and Saturn.[19]

Jupiter Highlights
• First spacecraft flybys of Jupiter (Pioneer 10-1973; Pioneer 11-1974). • Confirmed Jupiter emits twice as much energy as it receives from Sun. • Detailed studies of general banded structure of Jovian atmosphere, including the Red Spot. • Detected huge Jovian magnetic field. • Discovered Jovian radiation belt (energetic protons). • Close-up observations of Galilean satellites (Europa, Io, Ganymede, and Callisto).

Table 4. Selected scientific highlights of the Pioneer missions to Jupiter and Saturn. (continued)

Saturn Highlights
• First spacecraft flyby (Pioneer "Pioneer Saturn"-1979). • Confirmed that (like Jupiter) Saturn radiates about twice as much energy as it receives from Sun. • Discovered magnetic field around Saturn. • Discovered Saturnian radiation belts. • Observed that rings consisted of particles several centimeters in diameter (possibly made of frozen water and other ices). • Two new rings discovered (called F and G Rings). • Several new moons discovered.

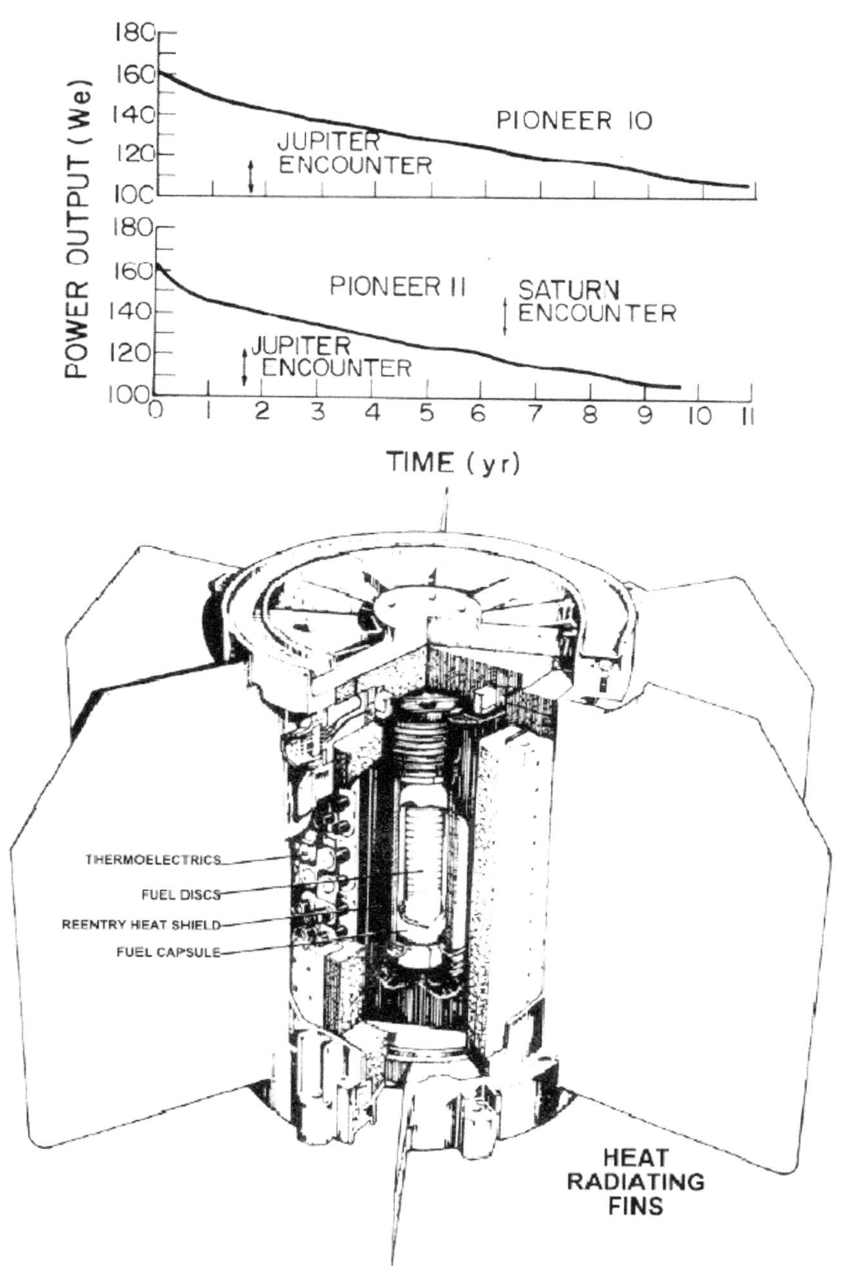

Fig.14. The SNAP-19 generator in the Pioneer program. Top is power output history for the Pioneer 10 and 11 spacecraft as of 1983 *From Bennett 1983.* Bottom is view of SNAP-19/Pioneer RTG. *Courtesy of U.S. Department of Energy.*

A series of U.S. missions to explore the planet Mars began in 1964 with Mariner 4 and continued with the Mariner 6 and 7 flybys in 1969 and the Mariner 9 orbital mission in 1971-1972. Project Viking, as designed, orbited Mars and landed a robot laboratory on its surface. Two identical spacecraft were launched from Cape Canaveral: Viking 1 on 20 August 1975 and Viking 2 on 9 September 1975. The landers were sterilized before launch to prevent contamination of Mars by terrestrial microorganisms. Viking 1 reached Mars orbit on 19 June 1976 and the first landing on Mars occurred on 20 July 1976 on the western slope of Chryse Planitia. Viking 2 began orbiting Mars on 7 August 1976 and its lander touched down on the Red Planet on 3 September 1976 at Utopia Planitia. Among their many outstanding scientific accomplishments, these nuclear powered robot laboratory landers discovered the following:[20]

- The highly oxidized Martian soil produced unique chemical reactions in the life detection equipment.
- The reddish color of the soil is due to oxidized iron.
- The winds at the surface of Mars are light, approximately 24 km / hr, while the surface temperatures range from about 189 K at night to 244 K in the afternoon.

Over 4,500 high quality images of the Martian landscape were returned by the two Viking landers. No evidence of the presence of living microorganisms was discovered in the soil near the landers. The question of life on Mars still remains open.

The unique mission requirements for the Pioneer and Viking missions created the need for variations in the SNAP-19 design. For example, the Viking landers had to withstand high temperature sterilization procedures in support of the planetary quarantine protocol, storage during the flight to Mars, and then the severe temperature extremes of the Martian surface; Pioneer had only to be designed for space operations. The design differences between the SNAP-19 generators used for the Pioneer spacecraft and for the Viking landers are shown in Table 5.

Table 5. SNAP-19 comparative design data. ***From Viking '75[21] and Goebel and Putnam 1979.***

	Pioneer	**Viking/Lander**
Design life (yr)	6	2
Overall Size (diameter x length)[1] (cm)	51 x 28	59 x 28(40)
Generator Weight (kg)	13.6	15.2
BOL Power Output (watts-electric)	41.2	42.5
Lead Voltage (volts)	4.0	4.4
BOL Fuel Inventory (watts-thermal)	648	683
BOL Pu-238 Inventory (curies)	19,500	20,600
Heat Shield Length (cm)	16.5	17.3
Capsule Length (cm)	12.2	12.7
Capsule Inner Liner Thickness (mm)	0.13	0.22
Capsule Weight (kg)	3.4	3.6
T/E Converter Initial Gas Fill (%)	75 He/25 Ar	90 He/I0Ar
Dome Reservoir Initial Gas Fill (%)	Not applicable	5 He/95 Ar
Capsule Cover Gas	Capsule exposed to T/E	Converter fill gas
Heat Shield Getter Recess	No	Yes
End Cover Attachment	Bolted/Seal-Weld	Lock Ring/Seal-Weld
Electrical Receptacle Seal	Single Viton O-Ring	Dual Viton O-Rings
Dome Reservoir-to-T/E Converter Seal	Not Applicable	Single Viton O-Ring
Hot Junction Temperature (K)	785	819
Number of Fins	6	6

[1] Diameter is across fin tips; length is mounting-flange-to mounting-flange except value in parenthesis for Viking is overall length including dome reservoir and pressure transducer boss on lower cover.

Details of the Viking/SNAP-19 RTG are shown in Fig. 15.

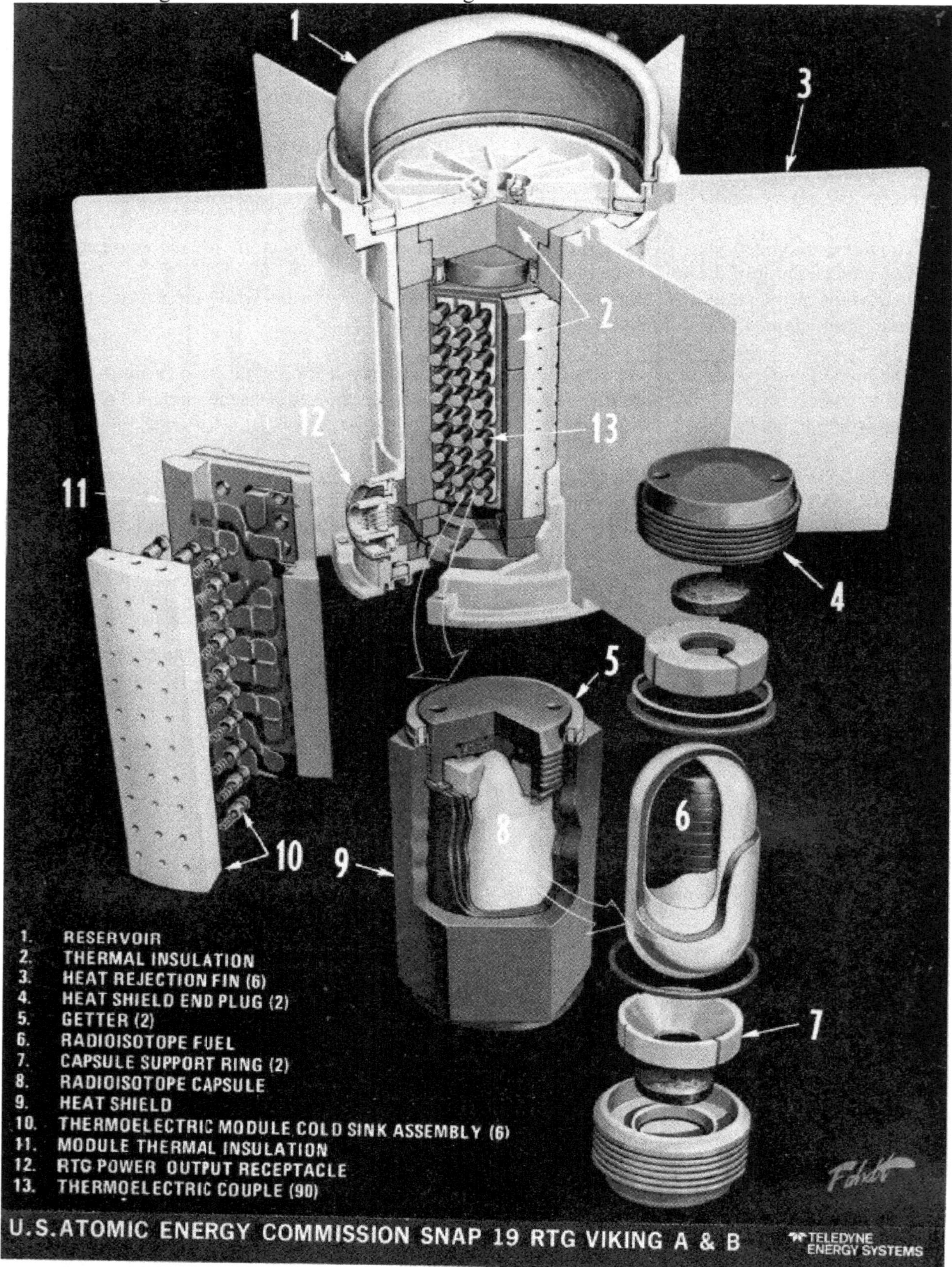

Fig. 15. Viking/SNAP-19 RTG. ***Courtesy of Teledyne Energy Systems.*** **contained within a hexagonal graphite shield.**

The heat source, (pictured in Fig. 16) contained plutonium-238 dioxide-molybdenum cermet hot-pressed into discs. These were arranged into a three layer tantalum alloy cylindrical fuel capsule The design provided the thermal energy and proper temperature distribution for the TE converters and also satisfied aerospace nuclear safety criteria. A four layer vent capsule configuration enclosed in a right hexagonal prism graphite body was used. Intact reentry through aerothermal heating and nuclear fuel retention in the event of impact was built into the design. The nuclear fuel provided approximately 682 watts-thermal at the beginning-of-life (BOL); each capsule containing 20,600 curies. The fuel capsule contained 18 or 19 plutonium discs and the initial power density of each disc was 3.2 W / cm^3. Other characteristic data concerning the Viking/SNAP-19 heat source are summarized in Table 5.

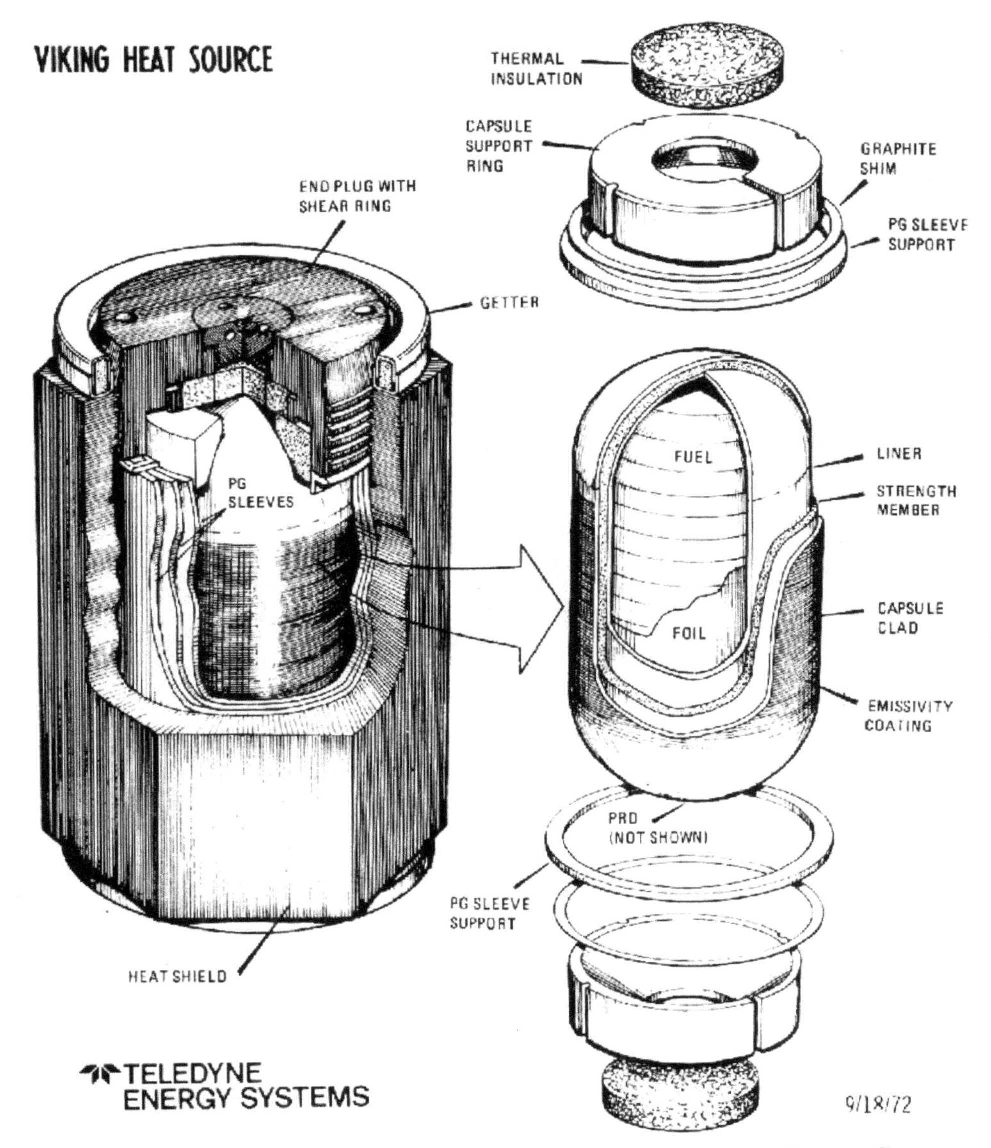

Fig. 16. Viking/SNAP-19 heat source. ***Courtesy of Teledyne Energy Systems.***

Between the fuel and other components there was incorporated an oxygen barrier. This used molybdenum-46% rhenium as the material. The primary fuel encapsulant used a 0.5 mm layer of tantalum-10% tungsten liner that facilitated easy decontamination during fuel assembly. In addition, a pressure relief valve was included to vent decay gases. A shell of tantalum alloy T-111 provided impact resistance. The final capsule shell, composed of platinum-20% rhodium, was used as an external oxygen barrier. In addition, a high emissivity coating of platinized alumina was placed on the capsule to facilitate heat transfer from the fuel capsule to the graphite heat shield. Graphite felt sleeves protected the ends from vibrations during launch. A zirconia support ring was used to relieve thermal stresses during reentry. Finally, a graphite heat shield surrounded the capsule and provided the ablation material needed during reentry and functioned as a heat sink during normal operation.

The thermoelectric converters consisted of TAGS thermoelectric material with a thin layer of SnTe at the hot side for the p-type material leg and PbTe for the n-type material leg. Each TE module contained 15 couples. Six modules were in the overall assembly. The couples were electrically connected in a parallel-series configuration. Fig. 17 illustrates details of the module arrangement. Six modules surrounded the hexagonally shaped heat source with a module resting on each of its six flats. A sheet of mica electrical insulation was used at the hot side between the individual modules and the heat source. Each module also had cold end hardware, including springs, pistons, alignment buttons, and heat sink bar. The TE converter also had thermal insulation to minimize thermal shut losses.

Conversion including springs, pistons, alignment buttons, and heat sink bar. Thermal insulation was used in the TE converter to minimize thermal shunt losses.

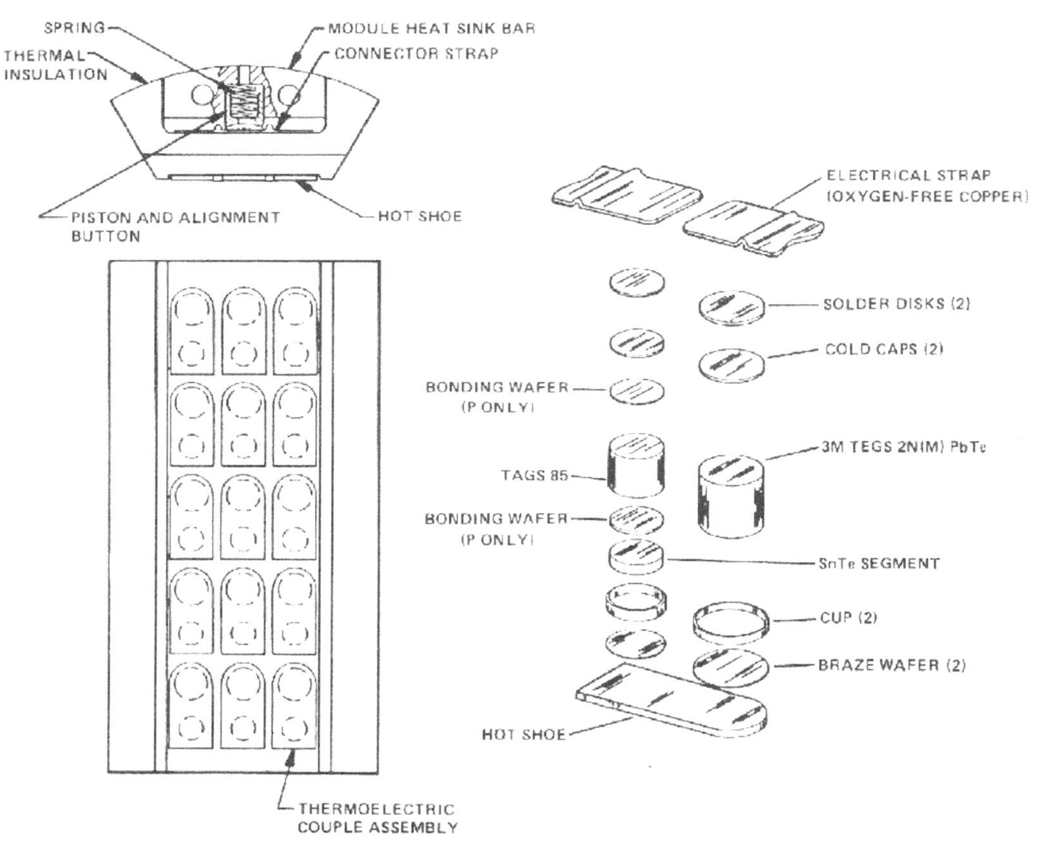

Fig. 17 Viking/SNAP-19 thermoelectric unit arrangement.[22]

A housing-radiator assembly consisted of a six-finned cylindrical housing, two ribbed end covers, and two seal-weld rings. The material for these components was magnesium-thorium alloy. In addition, a dome reservoir containing argon as a cover gas was used to retard sublimation of TE materials The thermoelectric converter and dome reservoir volumes were separated in the design in a way that permitted a controlled interchange of gases. The argon and the helium-rich TE cavity atmosphere were interchanged over time. This promoted increased thermal efficiency and enhanced electrical performance. The housing, fins, and dome radiator surfaces were coated with a zirconium oxide/sodium silicate emissivity coating. This coating material was attractive because it provided an emissivity greater than 0.83 with a solar absorptivity of only 0.2. The dimensions of the overall housing diameter across the fin tips was 58 cm and the overall height 29 cm. The power history of the Viking/SNAP-19 RTGs is given in Fig. 18.

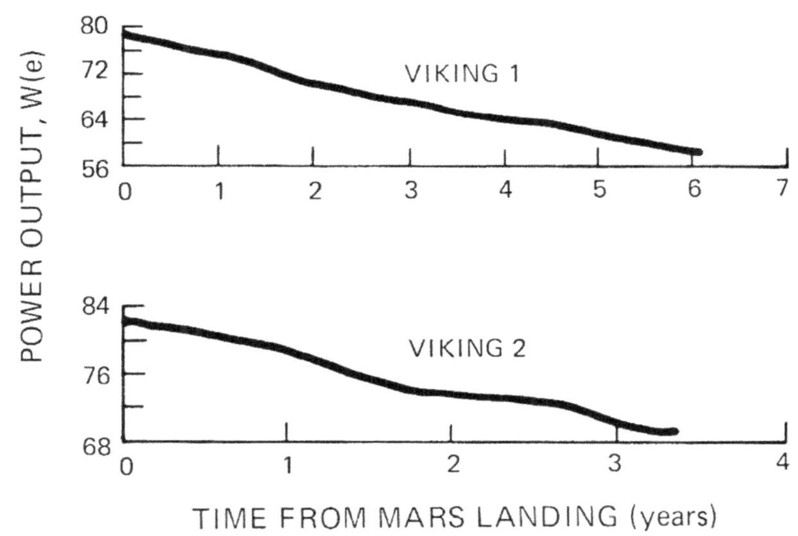

Fig. 18. Power history of Viking/SNAP-19 RTGs (smoothed data). ***From Bennett 1983..***

SNAP-27

The Apollo Program was the first time that nuclear power was integrated into a manned space program. The SNAP-27 system was used to power instrument and experiment packages deployed on the lunar surface.[23, 24] This required the incorporation of unique design features to the radioisotope-powered system as the astronauts were physically involved in its deployment (illustrated in Fig. 19). This included separate shipment packages for the radioisotope heat source and the TE generator. The major SNAP-27 hardware (see Fig. 20) included: (I) a generator with a thermopile, structure, and heat rejection system; (2) a fuel capsule that included the plutonium fuel in a separate hermetically sealed structure which prevented fuel release (see Fig. 21); and (3) a graphite lunar module fuel cask (GLFC) (see Fig. 22). The GLFC provided support for the fuel module during the journey to the Moon and thermal and blast protection in the event of a launch pad or mission abort. The aerospace nuclear safety criterion in effect during the Apollo Program was that, should there be an aborted mission, the probability of radiological exposure of the general public would be minimized.

The SNAP-27 fuel capsule and cask were designed to guarantee survival should a hazardous environment occur including launch pad explosion, an atmospheric ascent abort, and reentry into the Earth's atmosphere and ground impact. This was accomplished using a capsule with two fuel compartments integrated into a single high strength super alloy outer structure. Each fuel compartment contained $^{238}PuO_2$ fuel in the form of 50 to 250 micrometer-size 'microspheres' and produced a nominal 740 watts-thermal. The plutonium fuel was held in an annular configuration with void volume to accommodated long-term storage by confinement of the helium decay gas. The welded structure was 6.3 cm in diameter and approximately 41.9 cm long. A backplate helped secure the fuel capsule. The assembly weighed less than 6.8 kg. On the outer surface, a high emissivity iron titanate coating (greater than 0.85) was applied. The fuel capsule was qualified to meet thermal conditions as follows: (I) a steady state air temperature of 725 K; (2) ground storage at 530 K; (3) Earth-to-Moon transit in the GLFC at 1035 K; and (4) generator operation at 1005 K.

Fig. 19. The SNAP-27 heat source being removed from the Lunar Excursion Module (LEM) by Astronaut Gordon Bean during the Apollo 12 mission to the Moon in 1969. ***Courtesy of NASA.***

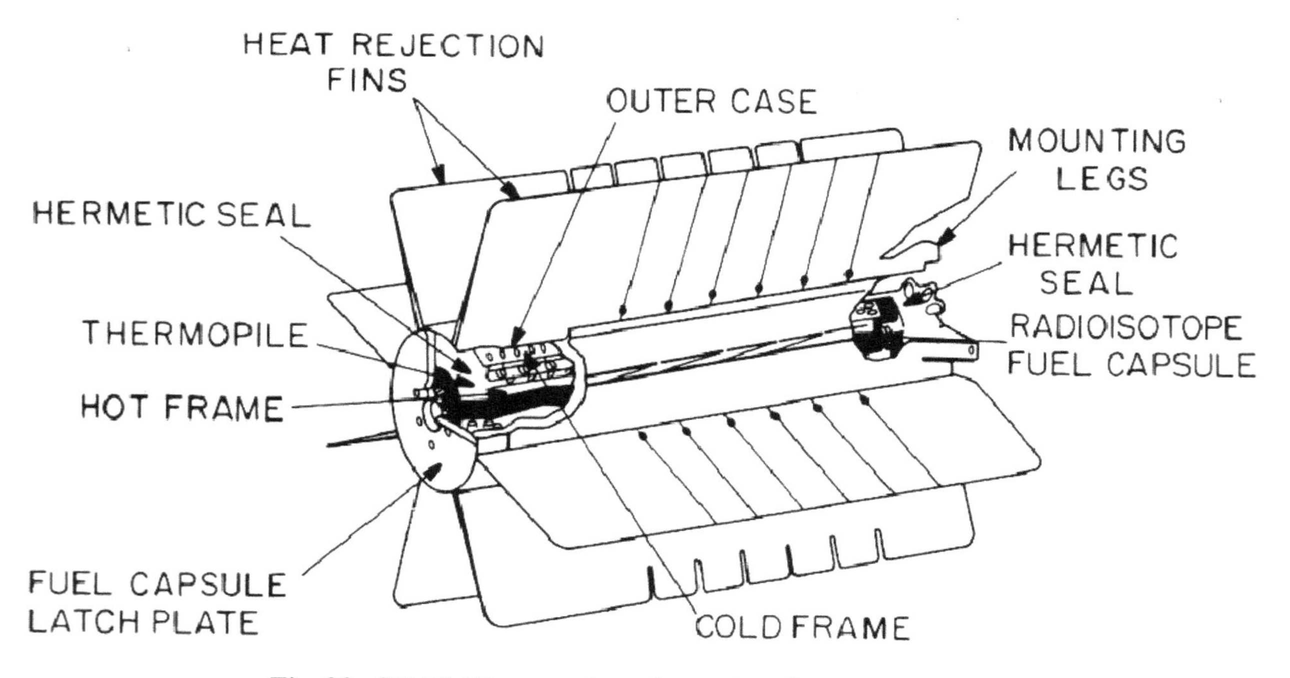

Fig. 20. SNAP-27 generator schematic. ***Courtesy of NASA***

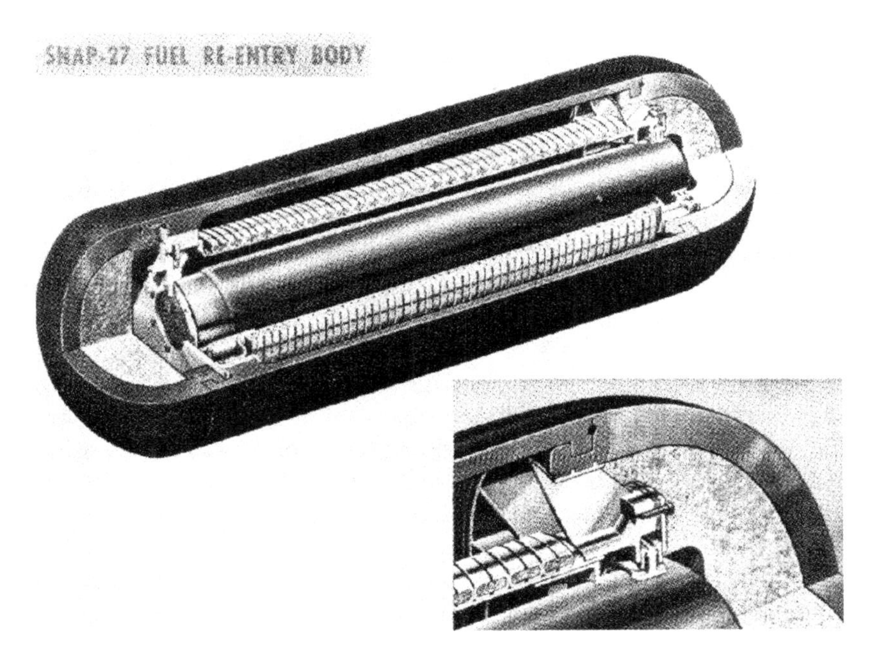

Fig. 21. SNAP-27 fuel capsule assembly. ***From Pitrolo, Rock, Remini and Leonard 1969.***

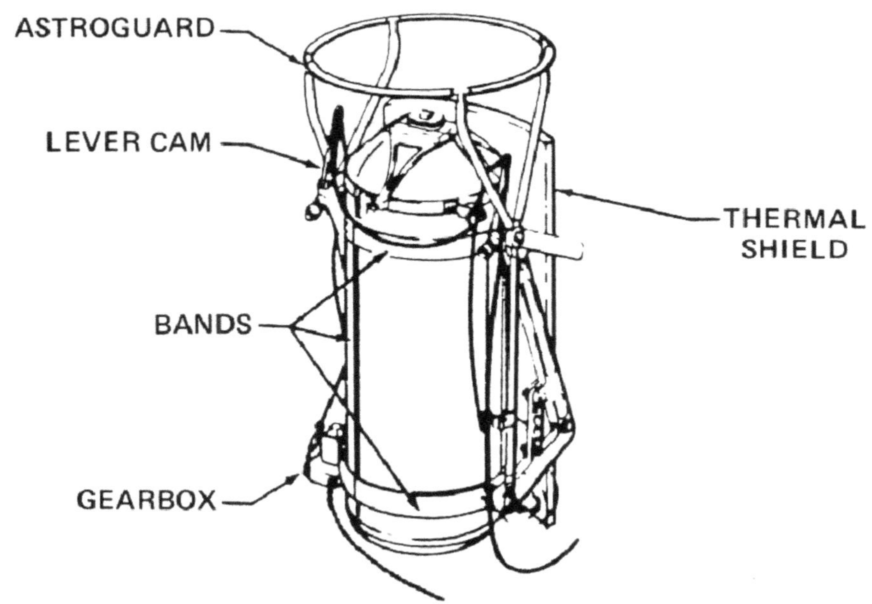

Fig. 22. SNAP-27 lunar fuel cask assembly. ***From Remini and Grayson 1970.***

The GLFC (see Fig 22) kept the fuel capsule together during the mission from launch to deployment on the lunar surface and served as an aerospace nuclear safety feature by enhancing containment if a reentry abort occurred. It consisted of a cylindrical container with a primary and secondary thermal shield and a support structure for the fuel capsule. The graphite primary heat shield formed the outer cylinder and hemispherical end caps. This protected the fuel capsule during reentry by acting as a heat sink, providing a radiation surface, and using ablation processes. A secondary thermal shield was also included. It consisted of a beryllium cylinder that was coated with successive layers of silver, rhodium, and a high emissivity coating (radifrax). This was used to promote radiant heat transfer and to provide oxidation resistance. In normal operation, heat from the fuel capsule was radiated to this beryllium cylinder, conducted through it, and then reradiated to the graphite cylinder. The beryllium cylinder was also designed to protect the fuel capsule in case of reentry by serving as a heat sink. By these means, a combination of rings, backplate, secondary thermal shield, and forward capsule support provided an independent internal structure to maintain fuel capsule integrity during postulated mission abort conditions.

Radiative coupling was used to transfer thermal energy from the fuel capsule to the generator hot frame. To promote radiation heat transfer, both surfaces were covered with a high emissivity iron titanate coating. The heat source surface had an emissivity of 0.85 and the hot frame surface an emissivity of 0.80. Temperatures when deployed on the lunar surface were 1,005 K for the fuel capsule and 880 K for the Inconel 102 alloy hot frame. The hot frame thermal energy was transported to the hot side of the TE converter materials by first passing through the electrical insulator, then through an electrode, and finally to a hot button. The hot junction temperature of the converter ranged between 855 K and 865 K, reflecting an overall temperature drop of 15 to 25 K. On the Moon's surface, temperatures can vary from 350 K during the lunar day to a frigid 100 K during the lunar night. The generator's cold side temperature operated at 545 K. In order to minimize heat losses, insulation was provided around the radiative coupler. Lead telluride (PbTe) served as the TE material and the couples were assembled in a series-parallel electrical arrangement. This prevented string loss, should a couple have failed open. Voltage output was between 14 and 16 volts DC. Each element was preloaded into its hot button by individual springs, sized and shimmed to establish a bearing pressure of 1.03 MPa. These couples then transferred thermal energy through beryllium oxide followers that interfaced with the cold cap. The massive beryllium frame was used to minimize temperature gradients. This outer case in addition functioned as the prime structural member of the generator.

The electric power histories of the SNAP-27 RTG units deployed on the lunar surface are presented in Fig. 23. The units functioned admirable and far exceeded the one year design life.

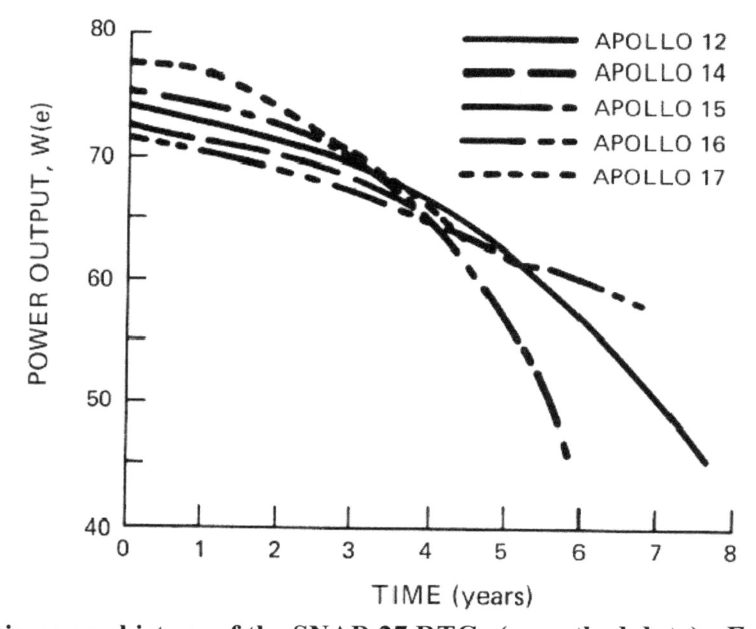

Fig. 23. Electric power history of the SNAP-27 RTGs (smoothed data). ***From Bennett 1983.***

Transit-RTG

The Transit-RTG was designed for use on navigational satellites and was the first radioisotope system developed for a five year life.[25, 26, 27] The beginning-of-life (BOL) power output was designed to be 34.2 watts-electric, with the unit weigh being less than 17 kg, and produce a minimum of 5 volts. TE converters of PbTe materials were incorporated into the unit. No cover gas was employed to inhibit sublimation.

This was the first generator to use radiative coupling between the heat source and the TE converters, although this was accomplished at some loss in efficiency.

The Transit-RTG was composed of the heat source and the converter as its major parts (see Fig. 24). The heat source, shown in Fig. 25, produced 850 watts-thermal and had a power density of 3.2 W / cm^3. The fuel was a plutonium dioxide molybdenum cermet with 82.5 wt % $Pu0_2$ and 17.5 wt % Mo. It was pressed into 5.4 cm in diameter discs that were 0.54 cm thick. An overcoat was not used on these discs. Twenty-two such fuel discs were placed in the 850 W heat source. The overall length of the heat source was 18.1 cm and an overall diameter of 12.7 cm. Approximately 2.64 kg of radioisotope fuel (some 25,500 curies) was used. The radiation levels generated in the vicinity of the heat source is shown in Fig. 26..

(a) Photograph of Transit 4A

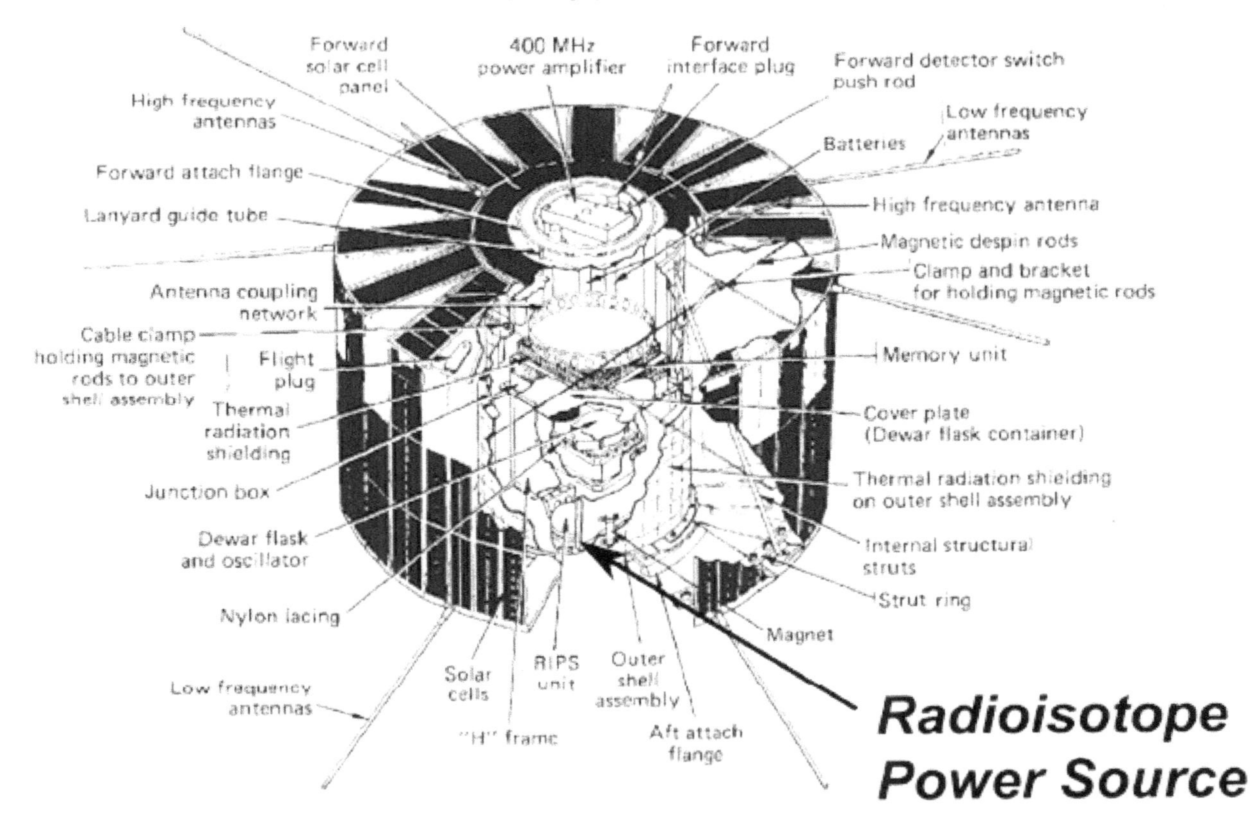

(b) Transit 4A Schematic Showing Internal Radiosiotope Power Source Location.

Fig. 24. Transit generator. ***Courtesy of TRW.***

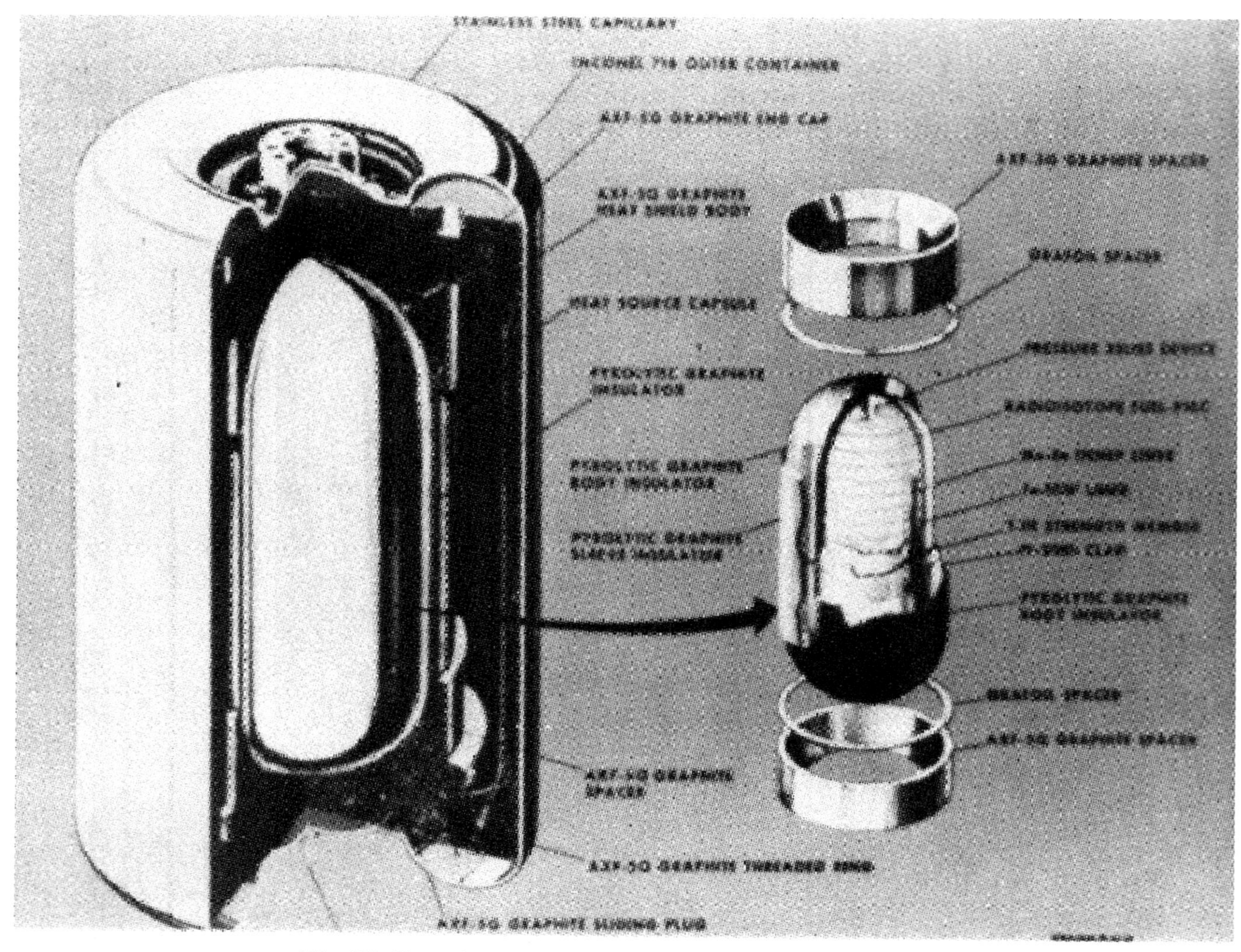

Fig. 25. Transit heat source design. ***Courtesy of TRW.***

The fuel capsule shown in Fig. 25 contained three subassemblies: (I) a three layer refractory metal capsule that included the fuel; (2) a two-layer graphite heat shield that resisted reentry environments and reduced the impact velocity of the heat source; and (3) an outer can that provided oxidation resistance during storage and prelaunch operations for up to 1,000 hours. Also included was a pressure relief device vent through a 4.6 m long capillary tube. Support for the fuel capsule was provided at the hemispherical ends. A preload was applied to the two graphite plugs to prevent vibrations during launch. A representative temperature distribution for the fuel capsule is shown in Fig. 27.

The thermoelectric converter was divided into twelve panels arranged into a regular polygon (see Fig. 28). The top closure plate and the bottom support plate did not contain TE converters and were insulated to minimize heat losses. A prism was formed by the twelve TE panels, 61.0 cm across the flats. The generator was attached to the spacecraft by means of a ring 34.3 cm in diameter. Each of the twelve TE panels consisted of 36 PbTe thermocouples interconnected in an electrical series-parallel network. To reduce sublimation of the TE material and to prevent the TE elements from short circuiting with the thermal insulation materials, each TE element was surrounded with molybdenum opacified quartz washers. The space between the heat collecting surfaces and the cold junction radiator was filled with a close-fitted multilayer insulation of alternate layers of aluminum foil and opacified quartz paper.

The Transit power plant weighed 12.8 kg composed of the converter weighing 5.8 kg, the heat source 6.3 kg, the support structure and cage 0.7 kg, and other miscellaneous components 0.11 kg.

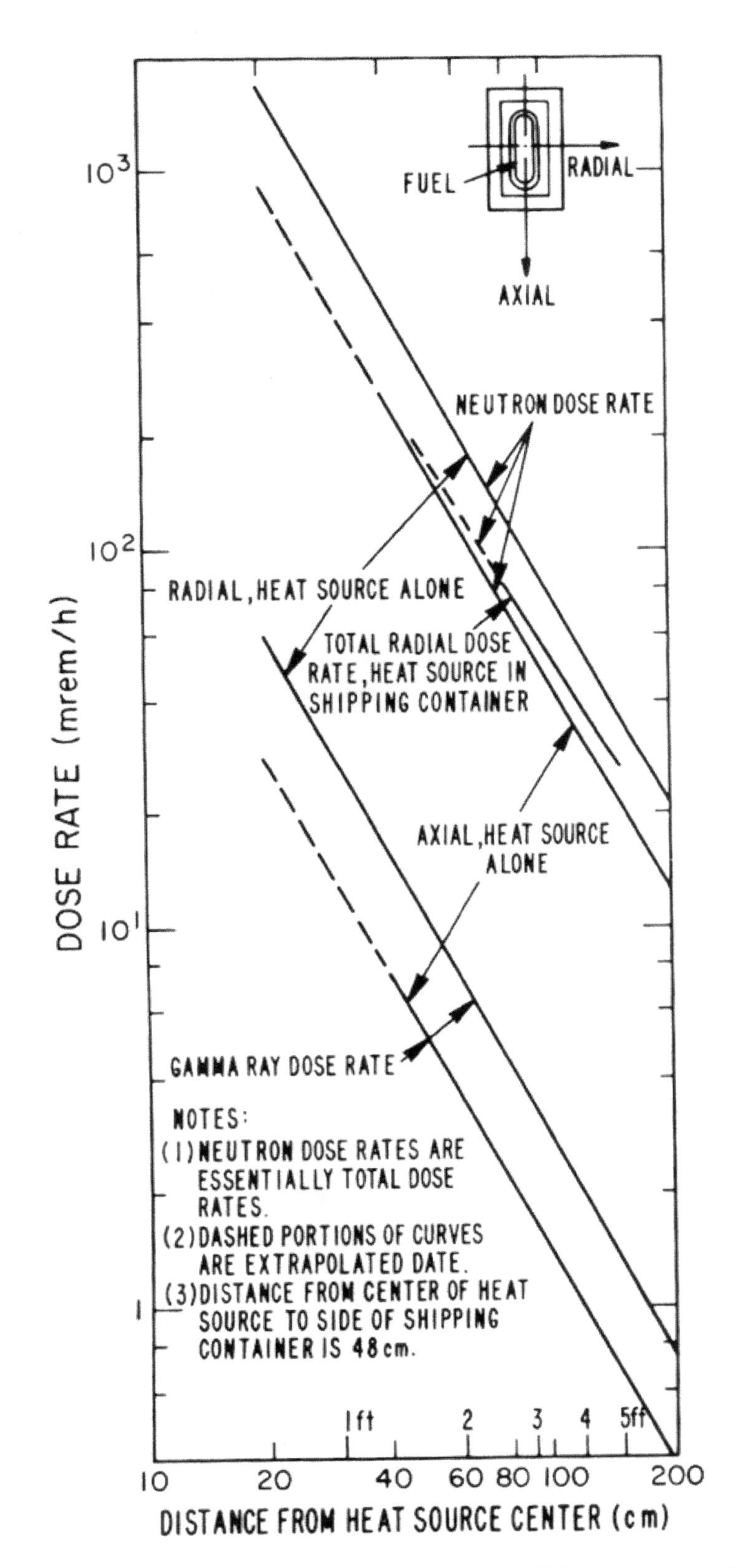

Fig. 26. Radiation levels as a function of distance from the Transit heat source. ***Courtesy of TRW.***

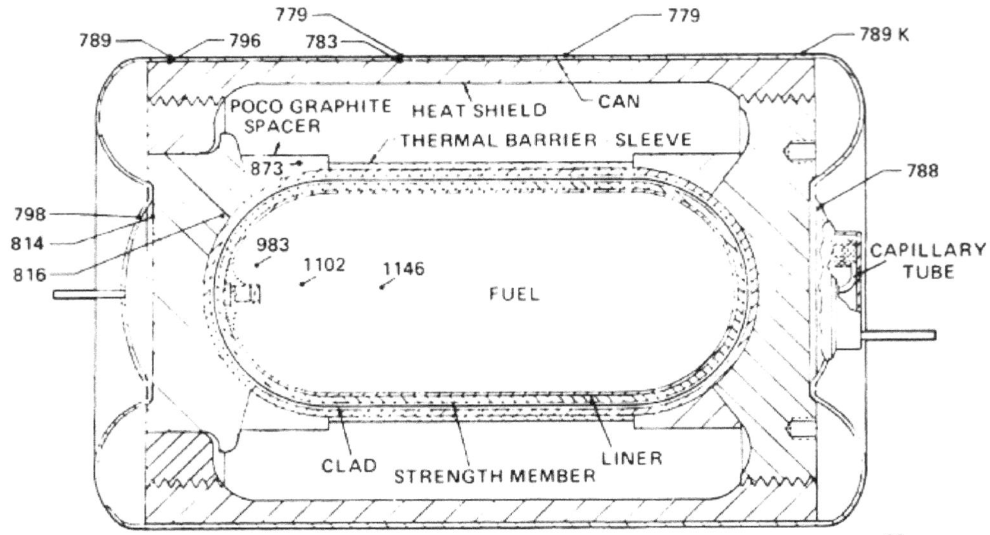

Fig. 27. Transit heat source temperatures (K) for a converter temperature of 673 K [28]

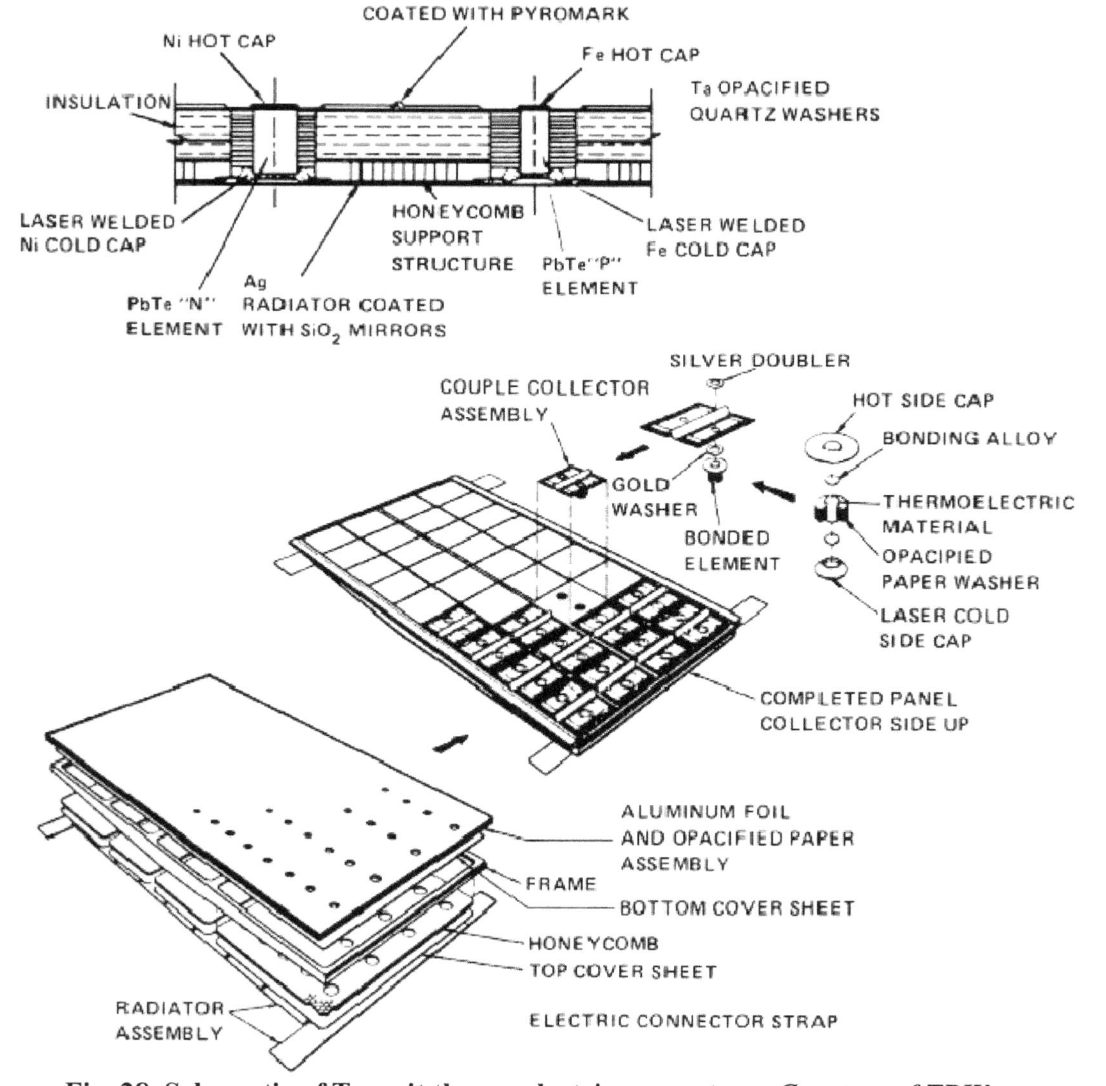

Fig. 28. Schematic of Transit thermoelectric converters. ***Courtesy of TRW.***

Multihundred Watt (MHW) Generator

A new generation of higher temperature generators was developed called the Multihundred Watt (MHW) generators. These higher temperature generators increased the heat source temperature as a result of a switch to silicon germanium (SiGe) as the thermoelectric material. This represented a major step forward in RTG design.[29, 30 31. 32 ,33, 34] Each MHW generator produced 150 watts-electric with an efficiency of 6.7 percent. LES 8/9 communications satellites each used two MHW generators and the Voyager spacecraft each used three MHW generators. The Voyager missions performed spectacular encounters with the giant outer planets Jupiter and Saturn. The encounters with Saturn occurred over four years after launch with all RTG units working as anticipated. The Voyager 2 spacecraft encountered Uranus in 1986 and Neptune in 1989.

The aerospace nuclear safety requirement that governed the design of the MHW heat source had immobilization of the plutonium fuel as the primary objective. This design philosophy supported the following objectives: (1) prohibit the release of nuclear fuel, especially in the form of biologically significant respirable particles; (2) minimize environmental (biospheric) contamination, particularly over populated land masses; and (3) maximize long-term immobilization following potential heat source accidents.

The MHW generator, shown in Fig. 29, major components were the radioisotope heat source and the converter. Performance data for the MHW design is summarized in Table 6. The heat source (see Fig. 30) was a right circular cylinder in shape with 24 fuel spheres. Each fuel sphere produced 100 watts-thermal energy. The fuel spheres had individual impact protection arranged around a spherical ball of plutonium dioxide ($Pu0_2$). These fuel spheres have a metallic (iridium) shell that contains the radioisotope fuel and a graphite impact shell which provides the primary resistance to mechanical impact loads. The primary structural element was a graphite aeroshell. This also served as a secondary ablator for shallow angle and orbital decay re-entries. During reentry, a graphite ablation sleeve provided the primary heat protection to the isotope heat source.

Fig. 29. MHW radioisotope thermoelectric generator. ***Courtesy of General Electric Co.***

Table 6. MHW design performance. *From General Electric. Document No.77SDS4206 1977 and GEA-932608(8-74)1M.*

Power output (watts)	
Beginning of mission	150
Five years	135
Weight (kg)	36
Output voltage (volts)	30 ±5
Envelope (cm)	
Diameter	38
Length	61
Average hot junction temp (K)	1,000
Thermoelectric material	SiGe
Magnetic field	< 15 x 10^{-9} tesla (15 x 10^{-5} gauss at 1 meter)
On pad	30 day unattended capability
Ground storage life (yr)	at least 2

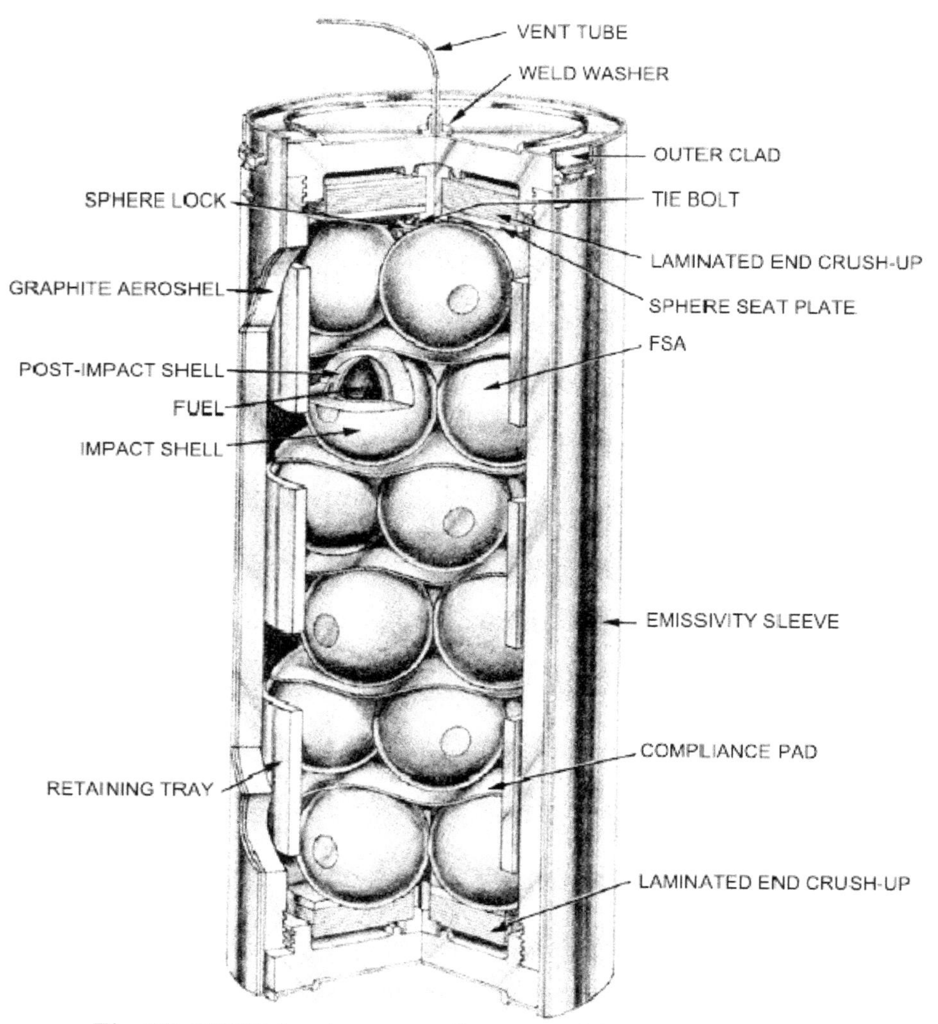

Fig. 30. MHW heat source. ***Courtesy of General Electric Co***

The individual fuel spheres are 3.7 cm in diameter with a power density of 3.6 W / cm^3. The 24 spheres contain a total of 6.05 kg of PuO_2. The activity level of 12.8 Ci / g corresponds to 77,300 curies per heat source. Fuel spheres are arranged in six planes of four spheres each. The total weight of the heat source is 19.6 kg; it has a surface temperature of 1,330 K. Since the plutonium-238 has a half-life of 87.75 years, the thermal power level of this heat source reduces approximately 1 percent a year.

The converter design contains a beryllium outer core and pressure domes. The TE unicouples are attached directly to the outer shell. Overall, the converter diameter is 39.7 cm and length 58.3 cm. The heat source is supported as a column within the converter by compression pads located at each end. Heat source loads are transmitted to the outer case through titanium end closure fittings that hold the compression pads and are attached internally to the outer case. The outer shell is coated with iron-titanate, with an emissivity of 0.85, to provide a heat rejection surface for the TE converters.

The power converter contains 312 SiGe unicouples arranged in 24 circumferential rows with each row containing 13 couples. Fig. 31 shows the unicouple design. These couples are made in unicouple assemblies that are individually bolted to the outer case. The radioisotope heat source thermal energy is transferred to the unicouple hot shoe by radiation. The hot shoe, made of SiMo (85 wt % Si) is bonded to the SiGe TE material. Two compositions of SiGe are used in the TE converter legs: 78 atom percent silicon for most of the length; and 63.5 atom per cent silicon for a short transition segment at the cold end of the couple to compensate for thermal expansion. For n-type TE material, the SiGe is doped with phosphorous; while for p-type material, boron is used as the dopant. The silicon alloy is bonded to a cold stack assembly of tungsten, copper, and alumina parts that separate the electrical and thermal currents. To retard silicon sublimation, the unicouple legs are coated with a one micrometer layer of silicon nitride (Si_3N_4). Heat losses are greatly reduced between the hot and cold shoes through the use of multi-foil insulation composed of 0.008 mm-thick molybdenum foil and SiO_2 cloth (Astroquartz) separators. Astroquartz yarn is also wound around the unicouple legs to electrically insulate the unicouples from the multi-foil insulation. The converter weighs 18.3 kg with the SiGe material making up 2.9 kg of this amount. The hot shoe temperature is 1,310 K and the cold strap is 575 K.

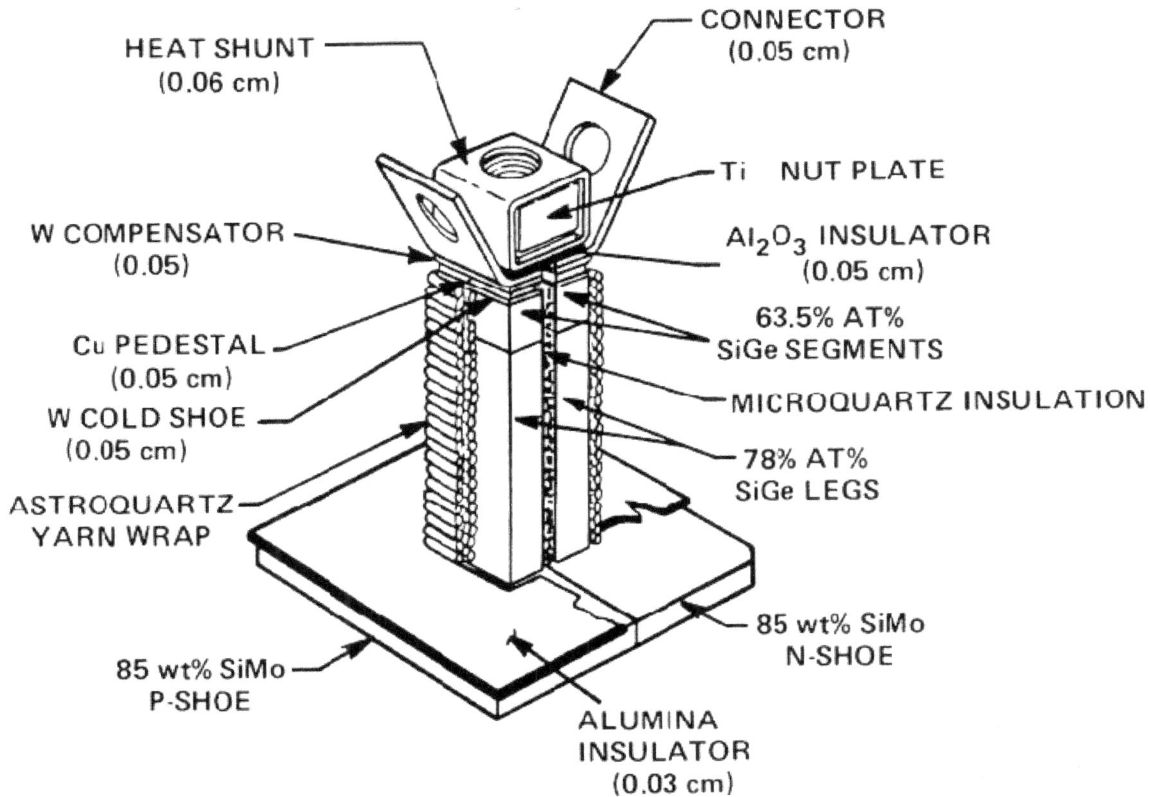

Fig. 31. SiGe unicouple for MHW generator. ***Courtesy of General Electric Co.***

Converter performance was extensively evaluated in both ground and flight tests Early testing identified several problems. One was silicon sublimation that was resolved by the addition of silicon nitride coating. Another problem was the top layer of the molybdenum foil breaking and shorting to the SiMo hot shoe. This was resolved by modifications in the design geometry and the use of an alumina spacer to separate the hot shoe from the multi foil insulation.

Long-term degradation mechanisms have also been identified for the MHW converters.[35] Dopant precipitation changes the TE properties of SiGe over time. These changes are attributed to a diffusion limited dopant precipitation process. Fig. 32 shows the changes in resistivity and Seebeck coefficient of the n-type TE material. Similar changes also occur in the p-type material, but in the 875 K to 1,275 K temperature range.

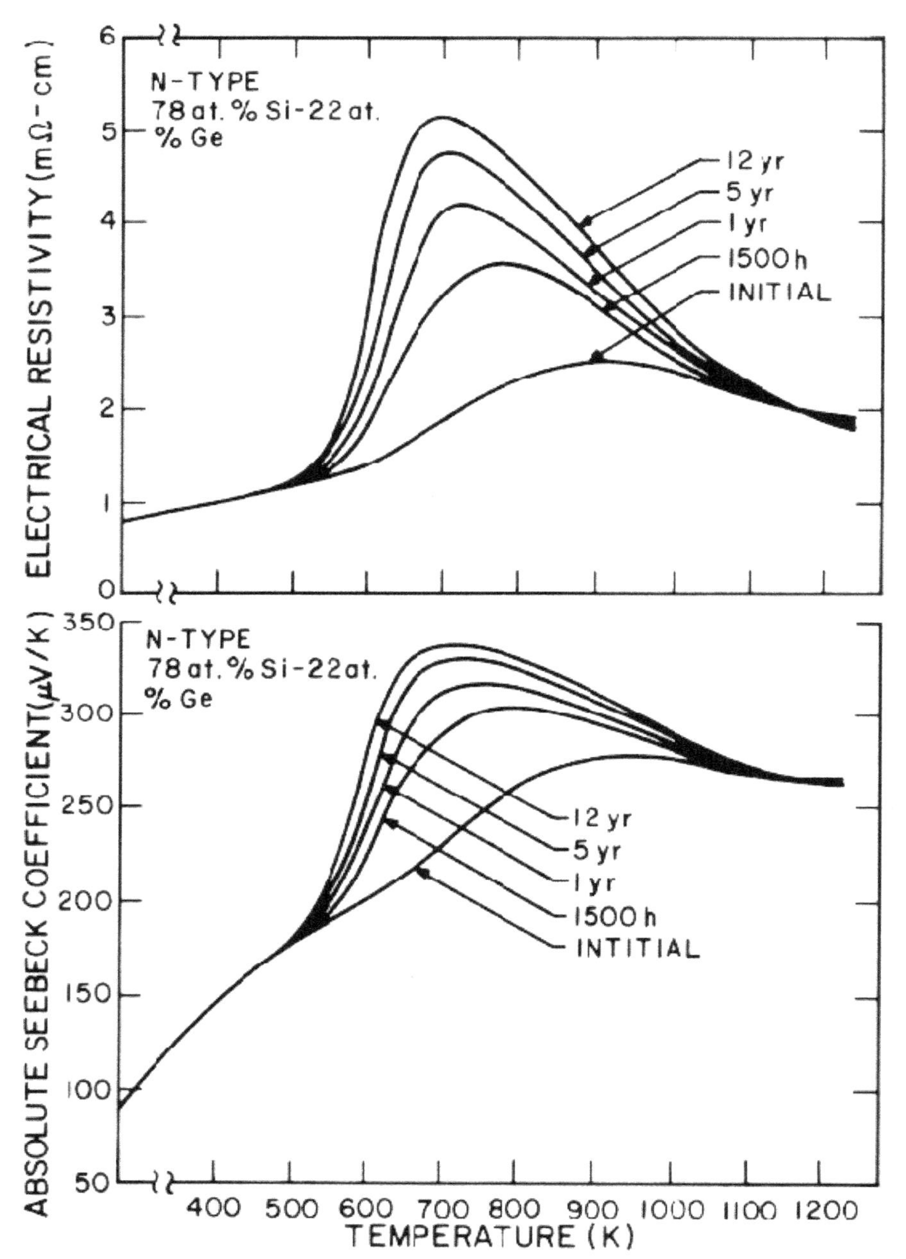

Fig. 32. Changes in the electrical resistivity and Seebeck coefficient of n-type SiGe thermoelectric material with time. ***From Kelly 1975.***

Material sublimation and reactions that can induce changes in thermal conductance and electrical shorting also cause performance degradation. At temperatures around the operating temperature of 1,275 K, some of the silicon reacts with the $Si0_2$ insulation to produce SiO vapor. The SiO vapor then diffuses to the cooler zones (875 K-1,075 K) of the thermopile and condenses. The condensing SiO tends to release free silicon and form $Si0_2$. Silicon and germanium also sublime and condense in the cooler zones. Thermal conductance is increased by the condensation of the leg wrap material. This also increases the emissivity of the molybdenum foils, resulting in higher heat losses and a decrease in power output. Partial shorts between the TE legs and foils were a problem before the Si_3N_4 coatings were applied.

A change in thermal conductance had been observed, due to an inhomogeneity in the fabricated TE materials. During operation, germanium-rich regions diffuse, resulting in lower thermal conductivity. The power loss for uncoated elements is typically 10 watts for a five year mission. However, this loss can be reduced to just 3 watts in five years using coated elements. Other long-term losses include: the thermal degradation of insulation, which contributes a power loss of about 3 watts in five years; and changes in foil electrical resistance of greater than 500 ohms, which produce a power loss of 0.15 to 0.6 watt.

Also, the effect of storage on these converters have been studied.[36] Two years of storage, followed by extensive testing of up to 32,500 hours, did not reveal any significant degradation. Such testing indicates that a projected life of 600,000 hours at nominal RTG temperatures can be anticipated, alleviating concern about the possible failure of SiGe TE generators due to structural degradation. LES 8/9 and Voyager I and 2 MHW flight generators demonstrated the successful operational use of these generators. Operational data is presented in Fig. 33 through 35.

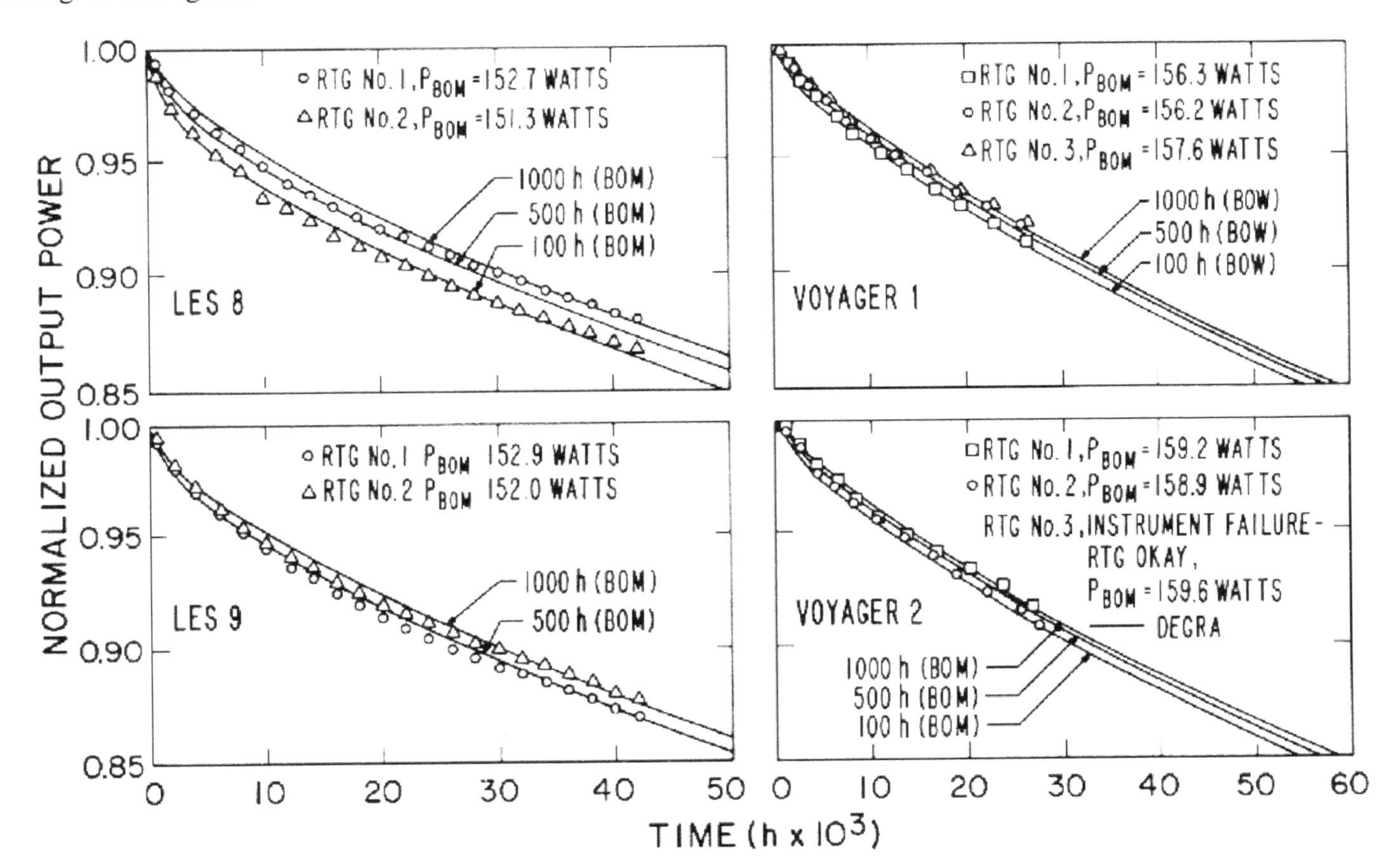

Fig. 33. MHW generator performance. ***From Shields 1981.***

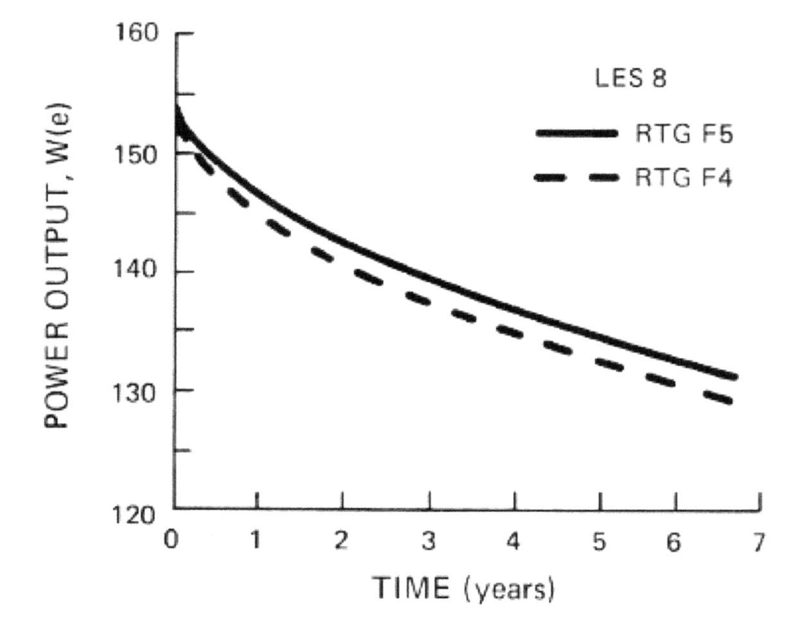

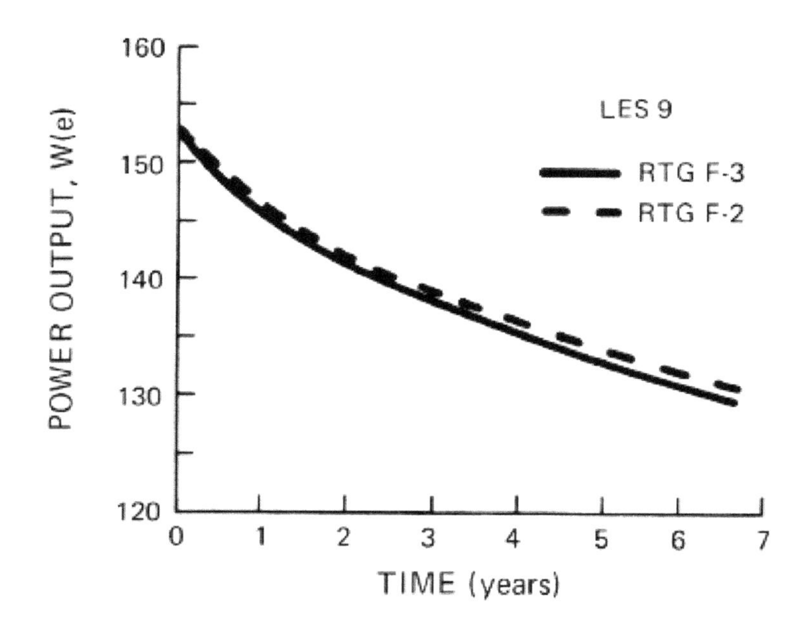

Fig. 34. Power history for LES 8 MHW RTG (top plot) and LES 9 MHW RTGs (bottom plot) (smoothed data), ***From Bennett 1983.***

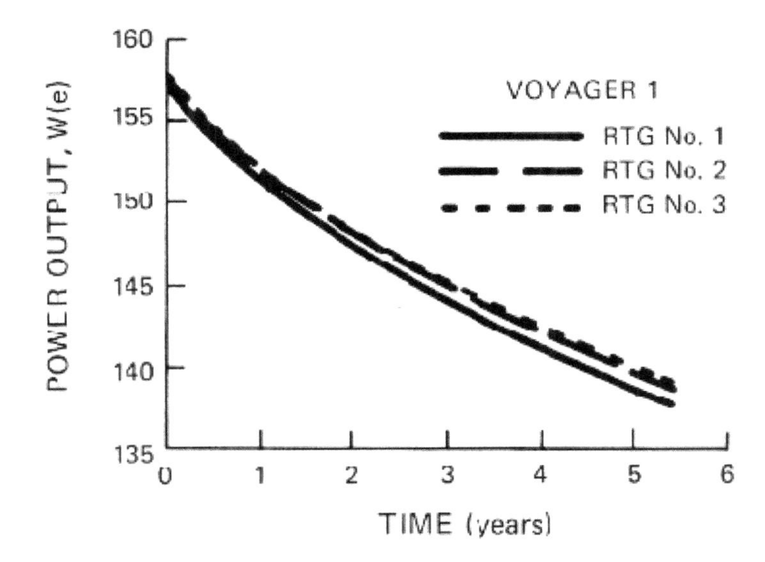

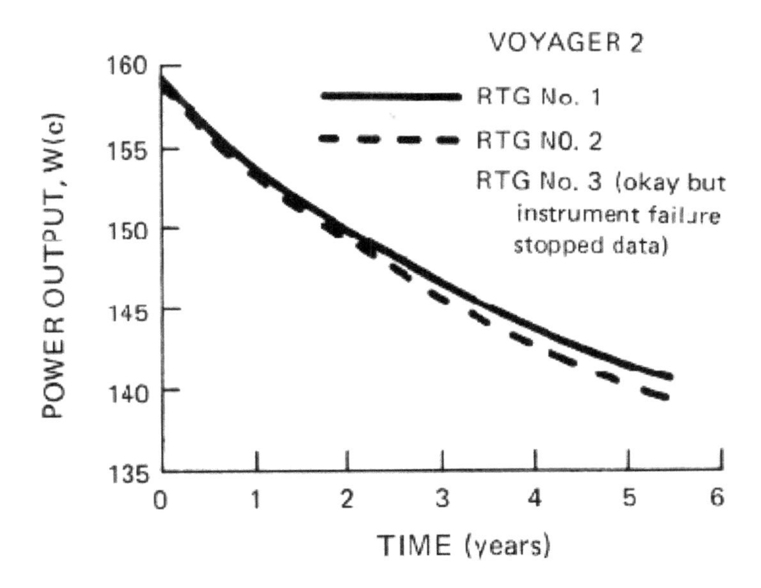

Fig. 35. Power history for Voyager 1 MHW RTGs (top) and Voyager 2 MHW RTGs (bottom) (smoothed data). ***From Bennett 1983.***

Summary

Table 7 presents a summary of the radioisotope generators developed by the United States for space power applications (GPHS-RTG and MMRTG will be discussed in the next chapter). Power levels for radioisotope generators have grown from the initial 2.7 watts-electric to 290 watts-electric for single units. For spacecraft requiring higher powers, multiple generator units have been successfully used. Specific power has also been increased from 1.3 w / kg to 5.2 w / kg--a factor of four. Efforts are underway to again double this performance.

There has been a continuous evolution in the fuel form from Pu metal to pressed PuO_2. The fuel modules are now a very mature modular design; the emphasis is now on improved power conversion technology and for designing systems for very severe environments. Especially challenging is the environments of the Mars surface and the environment of Venus.

Table 7. Radioisotope generator characteristics

	SNAP-3B	SNAP-9A	SNAP-19	SNAP-27	Transit-RTG	MHW	GPHS-RTG	MMRTG
Mission	Transit	Transit	Nimbus Pioneer Viking	Apollo	Transit	LES 8/9 Voyager	Galileo Ulysses Cassini New Horizons	Mars Science Laboratory
Fuel form	Pu metal	Pu metal	PuO_2-Mo cermet	PuO_2 microspheres	PuO_2-Mo cermet	Pressed PuO_2	Pressed PuO_2	Pressed PuO_2
Thermoelectric Material	PbTe	PbTe	PbTe-TAGS	PbSnTe	PbTe	SiGe	SiGe	PbTe-TAGS
BOL output power (We) (1)	2.7	26.8	28 - 43	63.5	36.8	150	290	120
Mass (kg)	2.1	12.2	13.6	30.8 (2)	13.5	38.5	54.4	43
Specific power (We/kg)	1.3	2.2	2.1 - 3.0	3.2 (3)	2.6	4.2	5.2	2.8
Conversion efficiency (%)	5.1	5.1	4.5 - 6.2	5	4.2	6.6	6.6	6
BOL fuel inventory (Wth)	52	565	645	1480	850	2400	~4400	~2000
Fuel quantity (curies)	1800	17000	34,400 - ~80,000	44500	25500	7.7×10^4	1.3×10^5	$\sim 5.9 \times 10^4$

(1) BOL = Beginning of life
(2) without cask
(3) includes 11.1 kg cask

Chapter 4

More Recent Radioisotope Generator Systems

In earlier missions, the radioisotope generators were designed for specific space missions or classes of missions. Higher power levels were sometimes met by using several RTGs on one mission. To meet an increased demand for power in new missions, one would have to add integral numbers of complete systems. For example, this was the approach used on the Voyager spacecraft that used three MMW generators on each spacecraft. However, this design approach often introduces weight and volume penalties[1]. Therefore, a modular power unit approach is now being pursued to provide design flexibility in meeting rising power demands for many missions. The General Purpose Heat Source (GPHS) [2, 3, 4] was developed as a modular heat source as a replacement for the Multihundred Watt (MHW) generator system. It has been successfully used on the Galileo, Ulysses, Cassini and New Horizons missions.

The latest radioisotope generator is the Multi-Mission Thermoelectric Generator (MMRTG). It is designed to provide 120 watts of electrical power from a heat source composed of eight General Purpose Heat Source (GPHS) modules. Its first application is the Mars Science Laboratory rover scheduled for launch in 2011 with a Mars surface landing in 2012. The rover is to operate at least one Mars year (687 Earth days).

General Purpose Heat Source-Radioisotope Thermoelectric Generator (GPHS-RTG [5, 6]

Design Configuration

The GPHS-RTG design, shown in Fig. 1, is a stacked column of 18 modules in the GPHS fuel assembly and 572 silicon-germanium (SiGe) alloy thermoelectric unicouples in the thermoelectric converter assembly. The overall generator diameter is 0.422 m and a length of 1.14 m with a mass of about 55.9 kg. Each GPHS module produces 245 W_t for a total of 4,410 W_t and the converters converts this into at least 285 W_e at beginning-of-mission. At the time of fueling, the GPHS-RTG is capable of producing more then 300 W_e. The total radiological inventory for a typical RTG is 10.9 kg of PuO_2 fuel (about 8.1 kg of Pu-238 per generator) with a total activity of about 132,500 curies. The GPHS-RTG has a specific power of 174.4 W_t / kg or 5.1 W_e / kg.

In case of an accident or launch abort situation, the fuel module size and shape were selected to survive reentry through the atmosphere and impact the Earth at a modest terminal velocity of 50 m / s. Each module, shown in Fig. 2, is a rectangular parallelpiped with the overall dimensions of 9.72 cm. x 9.32 cm. x 5.31 cm. Two graphite impact shells (GISs) are included in each module that operate at a nominal temperature of 1,335 K in space. The GIS units each contain two cylindrical fuel pellets 2.75 cm in diameter and length. Each fuel pellet is individually encapsulated in a welded iridium alloy containment shell or cladding with a minimum wall thickness of 0.05 cm. The iridium alloy is capable of resisting oxidation in a post-impact environment and provides chemical compatibility with the fuel and graphite components during high temperature operation and postulated accident conditions. A frit vent hole is provided to permit helium gas release from the radioisotope fuel pellet, but prevents a release of plutonia particulates. The combination of fuel pellet and cladding is called a fueled clad.

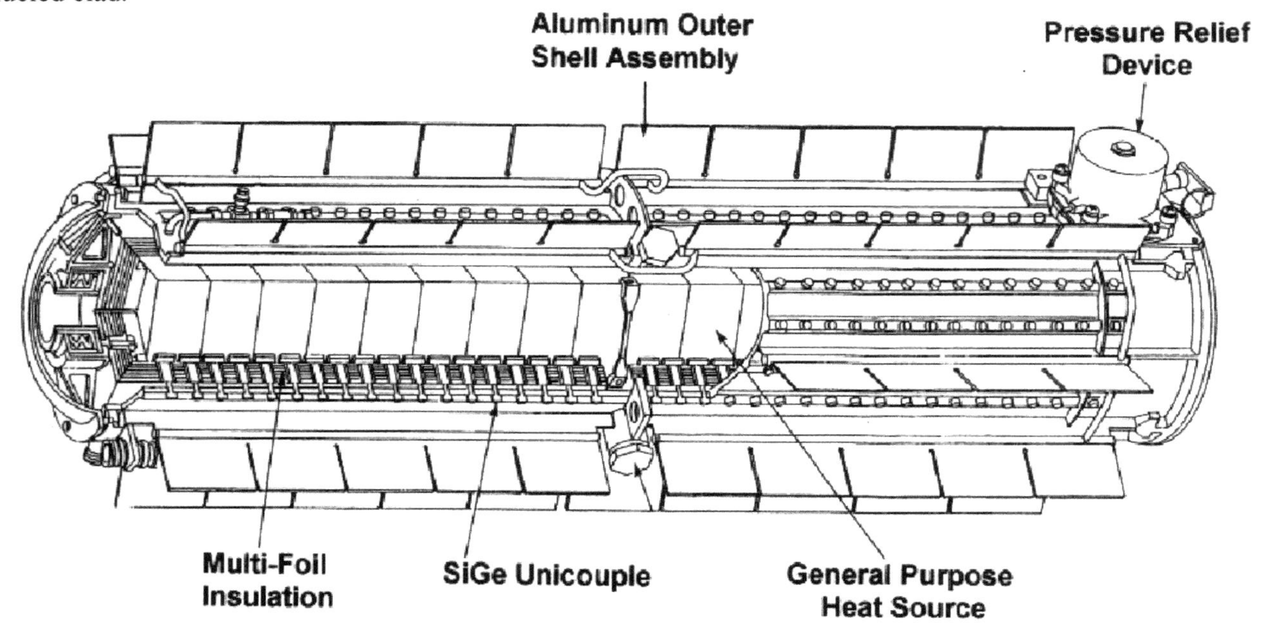

Fig. 1. General purpose heat source-RTG. *Courtesy of the U. S. Department of Energy and General Electric Co.*

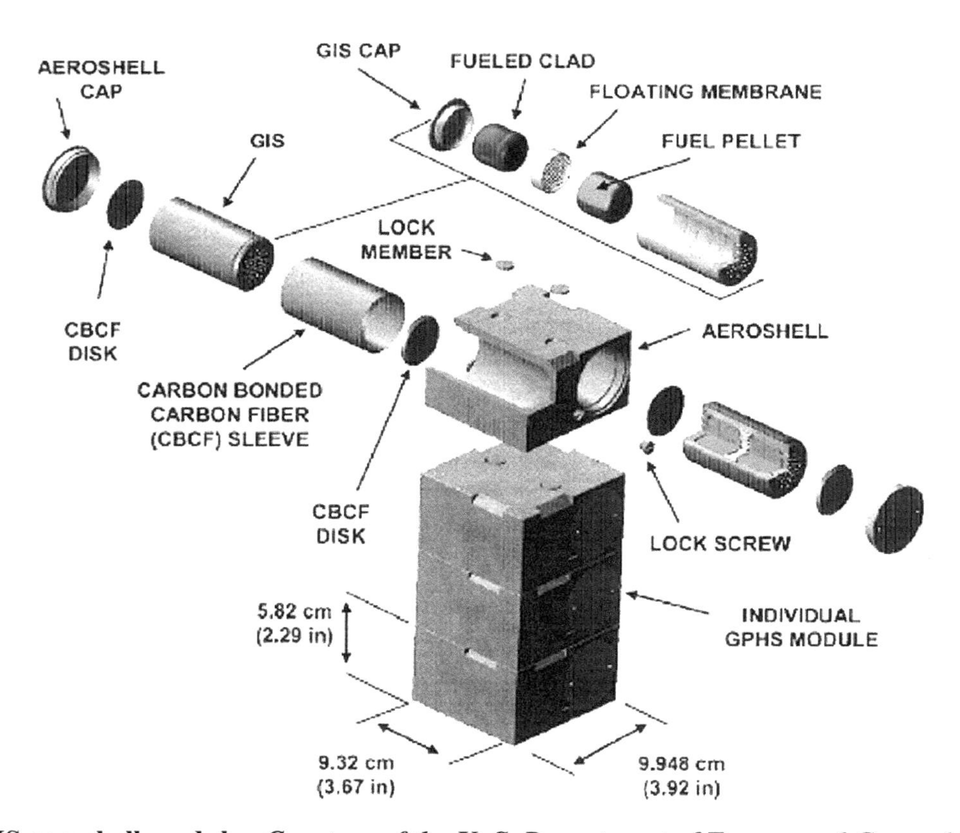

Fig. 2. GPHS aeroshell module. *Courtesy of the U. S. Department of Energy and General Electric Co.*

A fueled clad consists of a plutonium-238 fuel pellet (PuO_2) encased in an iridium shell that contains the fuel. Two of these fueled clads are combined in a graphite impact shell. These are separated by a floating graphite membrane that provides resistance to mechanical impact loads. The modules are constrained by locking members that minimize any relative lateral motion by individual modules. They are also packaged in a support system that provides axial compression to prevent separation of the modules.

The Graphite Impact Shell (GIS), with its two fueled clads, are made of Fine Weave Pierced Fabric (FWPF) carbon-carbon composite material. This design limits the damage to the iridium clads during free-fall or explosion fragment impacts. Two of these GISs are inserted into an aeroshell made also of FWPF graphite. To limit the peak temperature of the iridium cladding during atmospheric reentry heating and to maintain its ductility during the subsequent impact, a thermal insulation layer of Carbon Bonded Carbon Fiber (CBCF) graphite surrounds each GIS. The aeroshell is designed to contain the two GISs under severe reentry conditions; provide additional impact protection against hard surfaces at its terminal velocity; and provide protection for the fueled clads against over pressures and fragment impacts during postulated missile explosion events. The graphite aeroshell is the primary structural member to maintain the integrity and position of a stack of GPHS modules during normal operations, testing, transport, and launch.

The TE converter, the same as used in the MHW generator, surrounds the heat source; 572 TE couples convert the heat of radioactive decay directly into electricity. These SiGe unicouples operate at a hot side temperature of 1,273 K and a cold side reject temperature of 555 K. The corresponding nominal hot shoe temperature is about 1,308 K. The thermoelectric efficiency is about 6.6 per cent. An outer case provides the main support for the TE elements and the heat source. The heat source support system provides preload to enable the GPHS-RTG unit to withstand the mechanical stress environments of launch and ascent, and to avoid separation of the heat source modules during mission operation. The converter also provides axial and mid-span heat source supports, a multi-foil insulation packet and an internal frame, and the gas management system. A gas management valve in the pressure-tight outer case maintains the desired internal environment. An inert gas environment is used to permit partial power operation on the launch pad. This inert gas also protects the molybdenum foil and refractory materials during storage and ground operations. Once in space, the gas is vented by a pressure relief valve. The outer case is actively cooled up to the time the spacecraft is deployed.

The unicouples, shown in Fig. 3,[7] are individually fastened to the outer shell. Two silicon-germanium alloy legs of the couple with their corresponding sections of the silicon-molybdenum alloy (SiMo) hot shoe are doped to provide thermoelectric polarity. Phosphorous is the dopant for the N-leg and boron the dopant for the P-leg. Each leg is 2.74 mm x 6.50 mm in cross section, with a total length of 20.3 mm. The unicouple height is 31.1 mm and the hot shoe measures 22.9 mm and is 1.9 mm thick. The SiGe alloy thermocouple is bonded to a cold stack assembly of tungsten, copper, molybdenum, stainless steel, and alumina parts which separate the electrical and thermal currents. Copper connectors form the electrical circuit in the space between the inside of the outer shell converter housing and the outside of the insulation system. The unicouples are connected in two series-parallel electric wiring circuits in parallel to enhance reliability and provide the full output voltage. In the event of a single unicouple open circuit or short-circuit failure, power would still continue. The electric wiring is also arranged to minimize the magnetic field of the generator. Each unicouple is electrically insulated from the multi-foil insulation by several layers of Astroquartz (SiO_2) yarn (nominal diameter 0.76 mm) wound tightly around the couple legs and by an alumina wafer beneath the hot shoe.

The hot junction temperature averages about 1,273 K at BOM and the cold junction temperature averages about 560 K. The corresponding nominal hot shoe temperature is about 1,308 K.

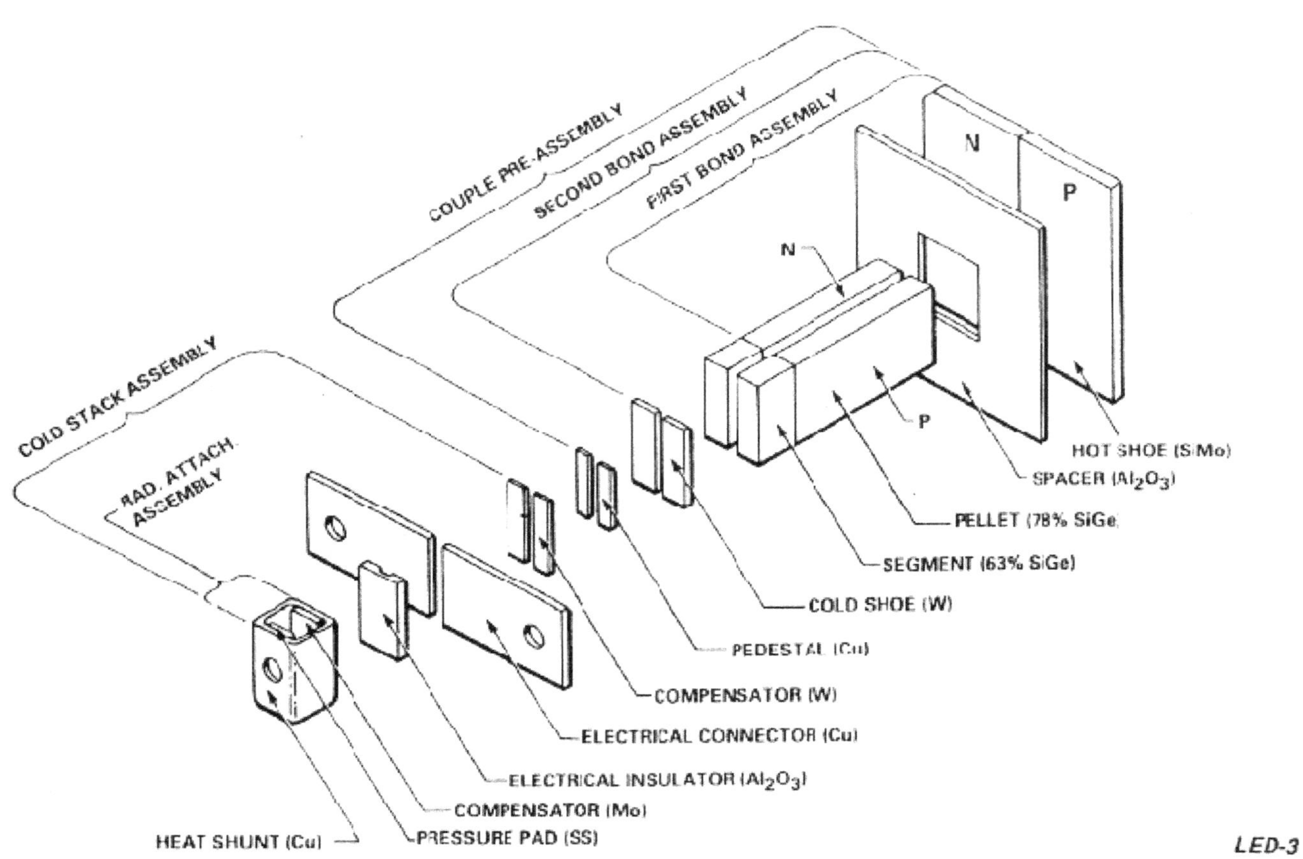

Fig. 3. Exploded view of a silicon-germanium alloy thermoelectric element ("unicouple") as used in the GPHS-RTGs and the MHW-RTGs. ***Courtesy of Lockheed Martin.***

Within the outer shell assembly are eight radial fins that provide the heat rejection to space for the generator units. The ribs are made of type 2219-T6 aluminum alloy forging with a silicone coating applied to the outer shell to raise the emissivity to 0.9. This is part of a flanged cylinder and four mid-span bosses made of the same aluminum alloy. Other components of the outer shell assembly include a multi-foil insulation assembly, which serves as a thermal barrier. This consists of 60 layers of molybdenum foil and 60 layers of Astroquartz. The support frame for the insulation system is made of molybdenum. Other components include the electrical power connector, four resistance temperature devices (RTDs), gas management system (GMS), and pressure relief device (PRD) are mounted to the outer shell and sealed using C-seals. On the inboard flange are mounted four barrel nuts on the four load carrying ribs to mount the GPHS-RTG to the spacecraft.

The GPHS-RTG is equipped with coolant channels through the fin roots to permit water circulation while installed in the shuttle payload bay. The auxiliary cooling is used prior to and during launch operations prior to the spacecraft being deployed. If necessary to return the payload to Earth, the auxiliary coolant system would also be used. The water circulation is designed to remove approximately 3,500 W_t. For expendable launch vehicles where conditioned gas streams can be used to cool the RTGs prior to launch, these coolant channels are not needed.

Table 1 summarizes the GPHS-RTG performance data.

Table 1. GPHS-RTG performance data.[8]

RTG Power Output (watts)	290 (BOM) 250 (minimum at EOM)
Operation Life (hr)	40,000 after launch
Weight (kg)	54.1
Output Voltage (volts)	28(+0.5)
Envelope (cm)	< 46 diameter x < 115 long
Hot Junction Temperature (K)	1275
Fuel	PPO (83.5 +1% $^{238}PuO_2$)
Thermoelectric Material	SiGe
Magnetic Field	$< 10 \times 10^{-9}$ tesla at one meter
On Pad	30 day unattended capability
Storage Life	1 year ground storage
Auxiliary Cooling	375 K avg. outer shell temp.

BOM = beginning of mission; EOM = end of mission.

Performance Tests

Testing philosophy required that the hardware be built and tested through increasing levels of assembly. Mission, spacecraft and launch vehicle requirements and environments were used to establish the top-level specifications. From these were derived the testing program to validate the GPHS-RTG system. The original test program was in support of the Ulysses mission (formerly the International Solar Polar Mission).

Thermoelectric unicouples were built and tested to verify that the MHW-RTG unicouple properties had been duplicated. Then modules of 18 unicouples were assembled and tested to validate the performance that the interconnected unicouples and associated hardware (e.g. insulation) function properly as a unit. For the Galileo/Ulysses RTG program, full-scale Component Engineering Test (CET) units were built and tested for structural and mass properties. The successful completion of such system level tests made the need to perform them for future missions unnecessary unless significant design changes were made.

For design verification, a non-nuclear electrically heated Engineering Unit (EU) was assembled and tested. This early test proved to be very important in that vibration test uncovered a problem that necessitated adding four clamps to hold the foil insulation basket. A nuclear heated Qualification Unit (QU) was used to validate the overall system design, assembly and performance.

The general sequence of tests performed on the flight units are summarized in Table 2.

Table 2. Performance sequence for GPHS-TRG assembly and testing.[9]

Converter fabrication and testing	LM[1] fabricated and processed the converter. Performance testing was done using an electrically heated thermoelectric generator (ETG) to measure power and other properties in vacuum and with an argon cover gas.
RTG fueling and processing	Mound [2] (later INL[3]) inserted the GPHS modules into the converter and measured the electrical performance both in a vacuum and with an argon cover gas.
Vibration testing	Mound (later INL) conducted these tests with the goal of determining the functional integrity (including resistance to leaks).
Magnetic field measurements	Mound measured the magnetic field.
Mass properties measurements	Mound (later INL) measured and/or calculated the masses, centers of mass, moments of inertia, etc. of the RTGs.
Nuclear radiation measurements	Mound (later INL) measured the neutron and gamma radiation dose rates of the RTGs.
Thermal vacuum tests	Mound (later INL) measured the RTG powers under simulated space (vacuum) conditions.

[1]LM = Lockheed-Martin Space Systems Company
[2]Mound = Mound Laboratory
[3]INL = Idaho National Laboratory

Each flight converter unit was first tested in a vacuum environment using an electrically heated source to provide a thermal simulation of a GPHS assembly. Table 3 shows the performance of each of the flight converters assemblies later used in the Galileo/ Ulysses/ Cassini/ New Horizons programs. The beginning-of-life (BOL) power, the power at the connector pins, was normalized to a thermal input of 4,402 W_t. The 'circuit isolation' resistance is a measure of the integrity of the electrical isolation or insulation system; it represents the insulation resistance between the thermoelectric circuit and the outer case. As the table shows, the isolation resistance was significantly greater than the 1,000 ohms requirement.

During the October 1983 testing, E-2 converter was accidently exposed to air for a brief period because of an imploded glove port. This led to some of the molybdenum foil surfaces having an increase in their emissivities and caused an electrical power loss of about 2.6 W_e from the values shown. E-2 was later used in the Cassini mission.

Table 3. Electrically heated thermoelectric generator (ETG) performance.[10]

Parameter	Requirement	E-1	E-2	E-3	E-4	E-5	E-6	E-7
Date		1/8/83	1/10/83	12/2/84	23/4/84	30/6/84	15/2/95	19/2/96
BOL Power (W_e)	293	295.6	294.8	298.1	296.0	297.5	293.4	294.6
Load Voltage (V)	30	30	30	30	30	30	30	30
Circuit Isolation (Ω)	>1,000	2,100	1,900	2,200	3,100	4,600	2,600	1,200

The radioisotope assemblies (qualification and flight units) were assembled in the Inert Atmosphere Assembly Chamber (IAAC) for initial functional performance measurements without the end domes. These measurements included power output, load voltage, open circuit voltage, current, internal resistance, isolation resistance, average outer case temperature (requirement less then or equal to 533 K) and bell jar temperature. Measurements were made in an argon atmosphere and also in vacuum to provide the initial data on the expected RTG performance.

Vibration tests were performed on the Engineering Unit, the Qualification Unit, and each of the flight units. The Engineering Unit and the Qualification Unit were subjected to flight acceptance (FA) and type acceptance (TA) vibrations. The type acceptance vibrations were 50% more severe in amplitude and longer in duration than the expected launch environment. Completion of the type acceptance vibration tests demonstrated that the GPHS-RTG design had more than sufficient structural strength margin. The tests were based on the dynamic environments encompassed by different mounting configurations used on the different spacecraft. The Engineering Unit successfully passed additional acoustic and pyrotechnic shock testing.

Magnetic performance criteria for the total dipolar magnetic field vector for the Galileo and Ulysses flight RTGs were 30 nT at 1 meter and 1 nT at 2 meters. Measurements yielded a maximum value of 148 nT at 1 m and an estimated 10 nT at 2 meters. The test criteria were waived for Galileo since the measured field was not a constraint on the mission. For the Ulysses RTG, compensation magnets were installed. The Cassini requirement was that the total dipolar magnetic field vector should not exceed 78 nT at 1 m from the geometric center of the RTG. The RTGs exceeded these limits. Separating the permanent magnetic field from the current-induced field, led to an acceptable results by the spacecraft customer.

Mass requirements on each flight system was that the RTG should be less than or equal to 56.2 kg. This was increased to 56.7 kg for Cassini and 58.0 kg for New Horizons. Table 4 shows that the mass criteria was met.

Table 4. Flight unit mass comparison[11]

Unit	Flight Mass (kilograms)
F-1 (Galileo)	55.95
F-2 (Cassini)	56.31
F-3 (Ulysses)	55.81
F-4 (Galileo)	55.92
F-5 (spare)	55.94

Table 4. Flight unit mass comparison (continued)

F-6 (Cassini)	56.45
F-7 (Cassini)	56.51
F-8 (New Horizons)	57.91

Nuclear Radiation specification for neutron emission rate from the unshielded GPHS assembly was not to exceed 7.0 x 10^3 neutrons per second per gram of plutonium-238, exclusive of any neutron multiplication obtained from the configuration of the fueled clads in the assembled heat source or attenuation within the RTG. An oxygen-16 exchange process was employed during the production of the fuel pellets to minimize the neutron emission rate. The Qualification Unit met specifications with a measured value of 5.9 x 10^3 neutrons per second per gram of plutonium-238.

Thermal vacuum tests simulate the space conditions at beginning-of-mission and end-of-mission for each RTG. These tests lasted from 6 to 40 hours in a thermal vacuum chamber at pressure of 0.1 mPa or less and an average sink temperature of about 309 K.

Safety Program[12, 13]

The GPHS-RTGs are designed to operate with a manned qualified launch vehicle--the Space Shuttle. The first missions to use these generators, the Galileo mission to Jupiter and the Ulysses mission to explore the polar regions of the Sun, were delayed by the Challenger accident in January 1986 for three years. The missions were redesigned with new upper stages. The Galileo launch configuration is shown in Fig. 4

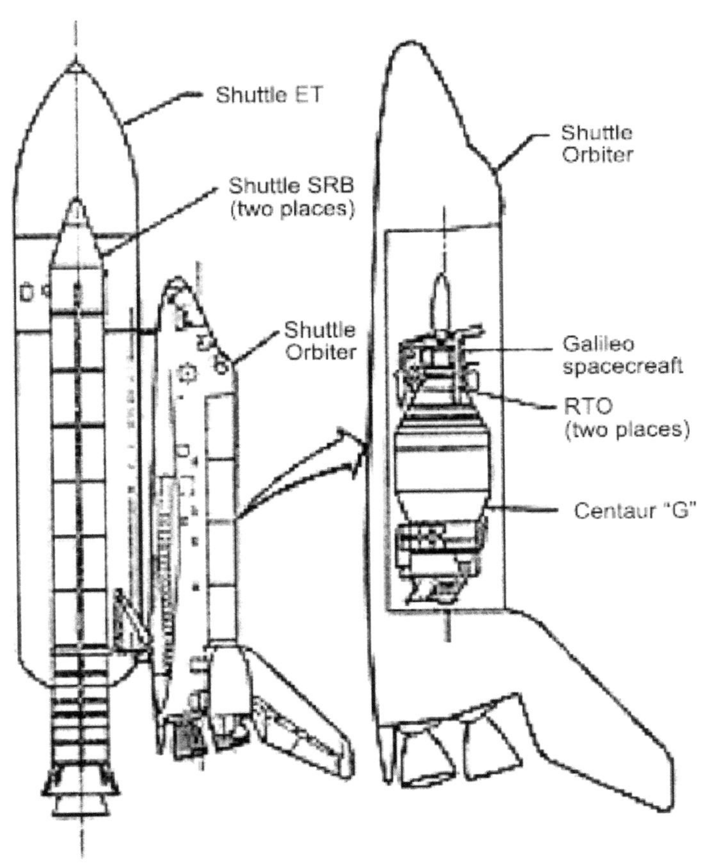

Fig. 4. Galileo launch configuration showing the position of the GPHS-RTGs in the Shuttle.

Figure 5 shows the various mission phases and provides the framework for performing the risk assessment. The phases are defined as follows:

Phase 0: Prelaunch/Launch This phase starts with the initiation of liquid propellants loading into the External Tank (ET) and ends with liftoff. The duration of the phase is from launch time T minus 8.0 h to T = 0 at launch.

Phase 1: Ascent This phase is from liftoff of the Space Shuttle/Inertial Upper Stage (IUS) vehicle from the launch pad at T = 0 and continues until the Solid Rocket Boosters (SRBs) are jettisoned at T +128 s.

Phase 2: Second Stage This phase includes the period from T = +128 s to T = + 532 s when the first burn of the orbital maneuvering system (OMS) engines begins. Included in Phase 2 are the events of the Shuttle main engine cutoff and the release of the External Tank.

Phase 3: On Orbit Phase 3 begins at T = + 532 s with the first burn of the Orbital Maneuvering System (OMS-1), includes the first and second burns of the OMSs for orbit attainment and circularization, opening of the cargo bay doors, the release of the IUS with spacecraft, and the firing of the Orbiter reaction control system to move the Orbiter away from the IUS/spacecraft and ends when the IUS with spacecraft is deployed from the Orbiter at T = +24,084 s.

Phase 4: Payload Deploy Phase 4 begins at T = +24,084 s and ends when the IUS has attained Earth escape velocity even though the spacecraft has not yet been deployed.

Phase 5: Venus Earth Earth Gravity Assist (VEEGA) Maneuver Phase 5 begins with the IUS attains Earth escape velocity. This phase includes the Venus flyby and the two Earth flybys. The phase ends when the second Earth flyby is completed successfully and the correct trajectory to Jupiter has been attained, which is approximately 3 years after launch.

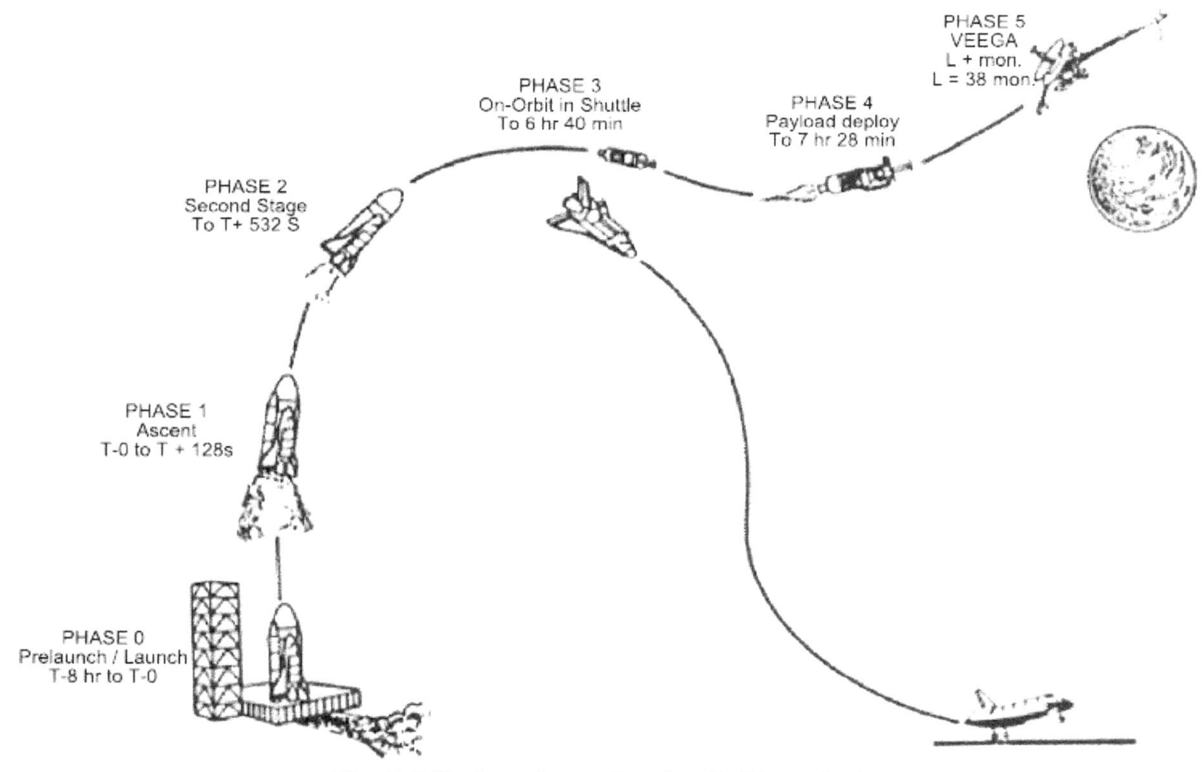

Fig. 5. Mission phases for the Galileo mission.

NASA developed a failure modes analysis of each phase as well as the environments produced by these failures and their probabilities of occurrence (See Fig. 6). These formed the basis of the safety test and accident analyses for the RTG-GPHS that were used in the Final Safety Analysis Report.

Bare fueled clad impacts tests were run in the original Galileo configuration and additional ones on the configuration after the Challenger accident for a total of 27 tests. These tests showed no release of fuel for impacts up to 250 m / s against sand and incipient fuel release in an impact of 117 m / s against concrete. To evaluate secondary impacts, bare fueled clads that had been damaged by solid rocket booster fragment impacts were impacted on concrete targets made of concrete cored from the Kennedy Space Center launch pad. Four bare fueled clads were impacted at velocities ranging from 45 to 60 m / s; the upper range of potential bare fueled clad impacts on the launch pad. In spite of severe damage, none of these fueled clads failed. These tests demonstrated that unbreached fueled clads damaged by solid rocket booster fragments are unlikely to fail in subsequent impact.

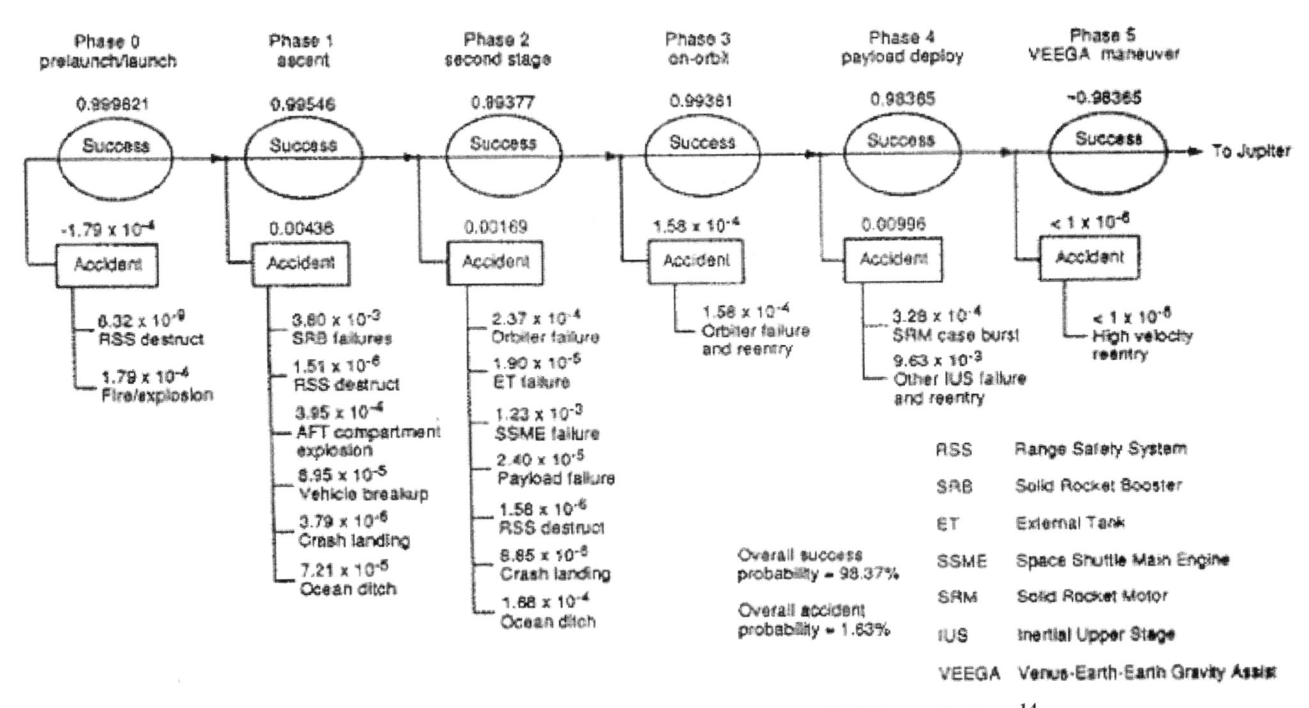

Fig. 6. Mission accident probabilities for the Galileo mission.[14]

Solid rocket booster (SRB) fragment tests determined the response of the GPHS fueled clads impacts by solid rocket booster fragments in a RTG-like configuration and to make a direct comparison between fueled clads containing fuel and that containing urania simulant. A gas-gun configuration was used (see Fig. 7). Five tests were conducted at velocities of 100 and 120 m / s. No fueled clad breaches occurred and no fuel released. Figure 8 shows that modeling data is in close agreement with the experimental results. The importance of having good modeling capability verified with experimental data cannot be over-emphasized when it comes to configuring future systems.

The large solid rocket booster fragment impact test series was to determine the response of the GPHS modules in the full RTG configuration to impact by a SRB fragment. Differences in response between modules near the end and modules in the central region of the RTG and between leading and trailing fueled clads were to be determined. No fueled clads were breached in face-on tests at velocities of 115 m / s and 212 m / s. A test on an edge-on configuration impacting at 95 m / s, the leading fueled clads were breached while the trailing fueled clads experienced only moderate deformation.

SRB fragment/Orbiter fuselage tests were to determine the velocity attenuation and the alteration of rotational motion of SRB fragments as a result of their interaction with the Orbiter structure prior to arrival of the fragments at the RTG location. A series of seven tests were run that supported the attenuation factors used in the revised FSAR. Also, the test data validated the models being used; thus it was possible to model a number of postulated adverse events without further test data. These included such events as: side-on and edge-on impact by SRB fragments; impact of intact RTG against the Shuttle Orbiter bay door; impact of a bare module against the Orbiter bay door; impact of a bare fueled clad against the Orbiter bay door; side-on impact of a bare GIS against concrete; and end-on impact of intact RTG on concrete, steel and wet sand. No clad failures resulted from face-on SRB fragment impacts and a small probability of fuel release from edge-on impacts.

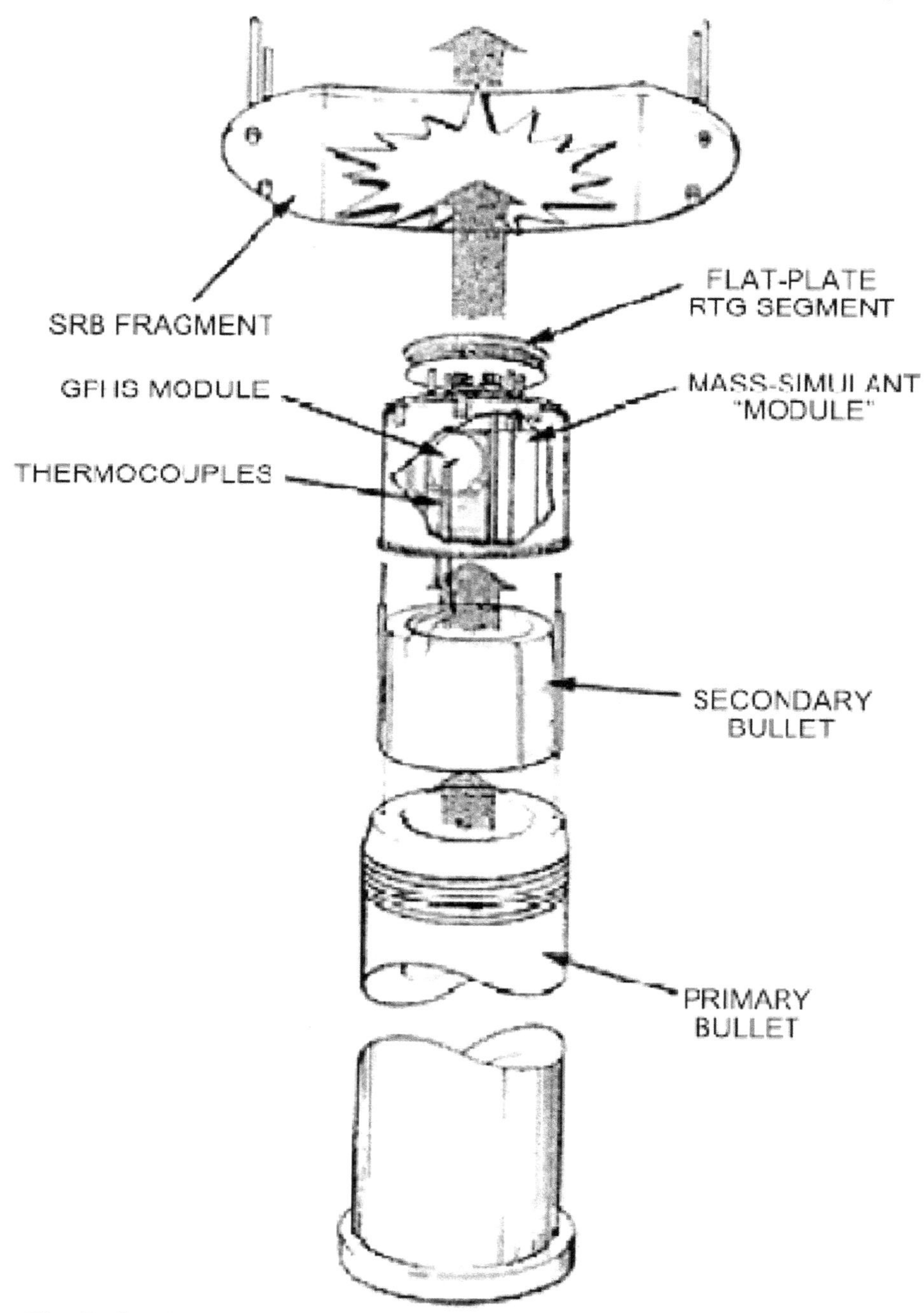

Fig. 7. Configuration for the gas-gun solid rocket booster (SRB) test.

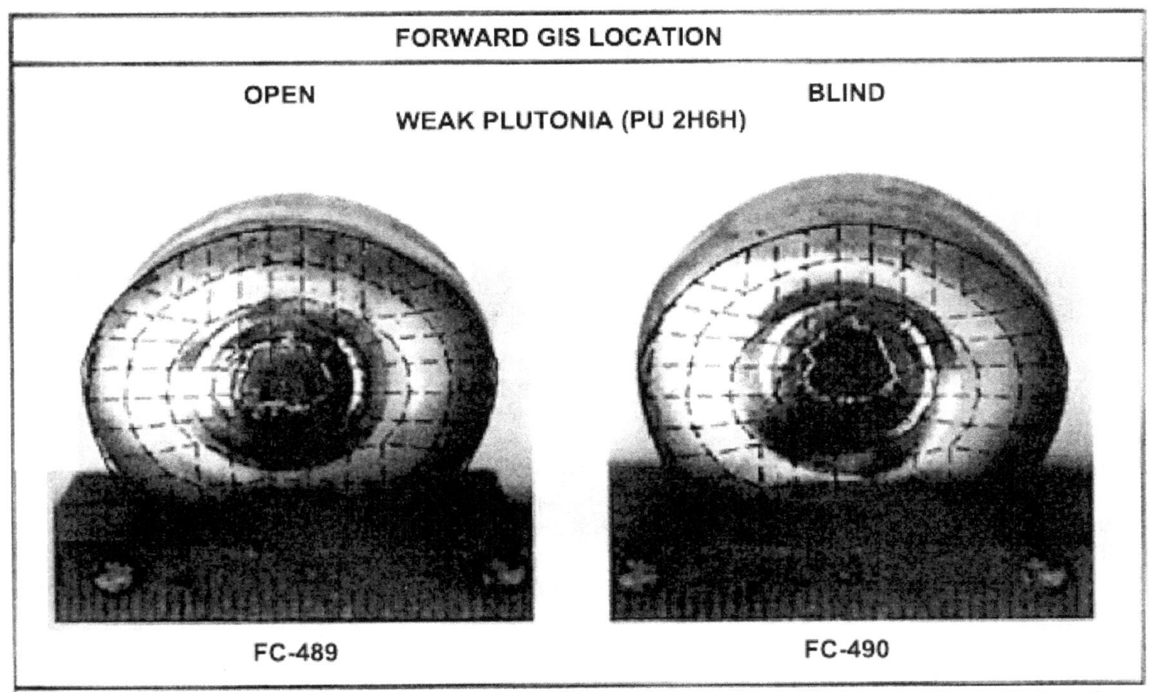

Fig. 8. Comparison of the predicted and observed post-impact geometries of the forward Graphite Impact Shell Plutonia-Fueled Clads from the Second Solid Rocket Booster (SRB) Fragment Test in the Gas Gun (FGT-2)[15]

Reentry assessment of the GPHS-RTG consisted primarily of analytical studies of the reentry conditions along with studies using arc jet tests. One scenario postulated accidents for a powered reentry of the IUS with the spacecraft. Earlier studies of orbital decay and the powered reentry of the Centaur upper stage bracketed the Galileo case. In these studies, the resulting aeroshell ablation was from 0.15 mm to 1.04 mm or 3.2% to 22.2% of the total minimum thickness. Thermal stress was not significant for these cases. Therefore, no further analysis was needed.

Postulated reentry of the Galileo spacecraft under potential conditions arising from malfunctions occurring during the VEEGA maneuvers were analyzed. Initial conditions include four steep (initial flight path of -90 degrees) and six shallow angle trajectories (initial flight path angles between -5 and -7 degrees). For the steep case, two initial speeds of 13.9 km / s and 15 km / s were used. The initial altitudes for the 13.9 km / s cases were 78.3 km and 84.7 km. For the 15 km / s cases the initial altitudes were 80.2 km and 86.6 km. In the shallow case, they used the same initial speeds with initial altitudes of 98.4 km, and 106.4 km for the 15 km / s and four initial altitudes from 95.1 km to 104.9 km for the 13.9 km / s initial speed. Analysis showed that the maximum recession of the aeroshell (3.0 mm) at the path angle of -5.53 degrees is well below the aeroshell minimum thickness of 4.7 mm. The iridium alloy clads temperature does not approach the melting point of iridium (2,727 K) nor does it approach the eutectic point with Fine-Weave Pierced Fabric (FWPF) (2,595 K). At the aeroshell temperatures (3,683 K to 4,000 K) calculated for the reentry, the FWPF is postulated to have limited strength. Graphite sublimation temperature is 4,100 K and the triple point of graphite is in the range from 5,000 K to 5,050 K at 10 MPa. However, graphite can soften at temperatures below its evaporation temperature. Because of the high temperatures and lack of data on mechanical properties at these high temperatures, a test program was conducted to assess the full-scale behavior of the GPHS module and the graphite impact shell (GIS).

The first tests were conducted in the 20-MW Aerodynamic heating Facility where moderate environments corresponding to an orbital decay reentry could be simulated. These test showed the conservatism of the analyze in that it took more ablation to release the GISs than the analyze showed. In fact, the threaded regions of the sidewall of the aeroshell have to be ablated to release the GISs. A second set of tests in a 60-MW

Interaction Heating Facility using arc jets allowed for significantly higher heating and pressure environments. The primary objectives of these tests were: (1) to determine the effect of the higher heating rates and pressures on the ablation response, including a planned test to failure; and (2) to gather sublimation data on the aeroshell as a check on the sublimation computer model. These test demonstrated that in a face-on, stable orientation the aeroshell will contain the GISs well beyond the analyzed release point. The macroscopic results of the aeroshell performance suggest a higher probability of reentry of the complete module assembly which means it is more likely that the full assembly will impact rather than free GISs.

Table 6 presents a brief summary of the GPHS-RTG safety test and analysis data base and is typical of the safety analysis performed for each mission using radioisotope generators.

Table 6. Overview of GPHS-RTG safety test and analysis data base.

Launch-Related Accident Environments
- No release of fuel at overpressures up to 15.2 MPa (fueled clads have capability beyond this) (Note: This overpressure is well beyond what could happen in a Shuttle accident.). - Threshold for direct mechanical failure of a GPHS fueled clad ~ 555 m / s with aluminum bullets and ~ 425 m / s with titanium bullets. - Minimal breaching fueled clad in collision with 3.53 mm thick aluminum flyer plate traveling at 1,170 m / s (beyond credible Shuttle flyer plate velocities). - No release of fuel in 210 m / s face-on impact of SRB fragment (limited breaching in less likely edge-on orientation at 95 m / s). (Note: This more than spans the range of credible SRB fragment impacts.) - No release of fuel for impacts of bare fueled clads at velocities up to and including 250 m / s against sand (incipient fuel release in an impact of 177 m / s against concrete.) (Note: At terminal velocity (~ 52m / s), the GPHS modules yield no release.) - No fuel release in fueled clads that had been impacted against KSC concrete at velocities ranging from 45 to 60 m / s (upper range) following impact by SRB fragments. - No release of fuel from the most severe liquid propellant and solid propellant fires.
Reentry-Related Accident Environments
- Modules survive orbital decay reentry (most probable of the reentries) and reentry at Earth escape velocity. - VEEGA reentry analysis yields no atmospheric release of fuel.
Environmental Properties
- Iridium alloy cladding resists fuel release for virtually unlimited periods in ocean water and in air. - Plutonia fuel is highly insoluble in water; there is limited migration of the fuel and limited uptake in plants.

Mission Behavior

Launch delays were experienced by the Galileo and Ulysses missions of three to four years as a result of the Challenger accident. The power requirements for Cassini and New Horizons were similarly adjusted to reflect their RTG fuel loadings and spacecraft launch dates.

Galileo GPHS-RTGs met the power requirements and enabled NASA and JPL to extend the mission three times until 2003. With the spacecraft running out of onboard propellant, the Galileo Orbiter was deliberately inserted into the Jupiter atmosphere on 21 September 2003. This prevented any possible impact on Europa. Scientist believe that Europa contains an ocean, which could mean a location for life. Galileo had successfully completed 35 orbits of Jupiter and provided a great wealth of data on Jupiter, the Jovian system and the four largest satellites of Io, Europa, Ganymede, and Callisto. The end was a dramatic insertion of the Galileo Probe into the Jovian atmosphere. Fig. 9 shows the performance of the RTGs from 1989 until 1997. The requirement of 568 W_e at the BOM was exceeded by a value of 577.2 W_e based on telemetry data. The power at EOM was 482 W_e which exceeded the specification requirement of 470 W_e by 12 W_e.

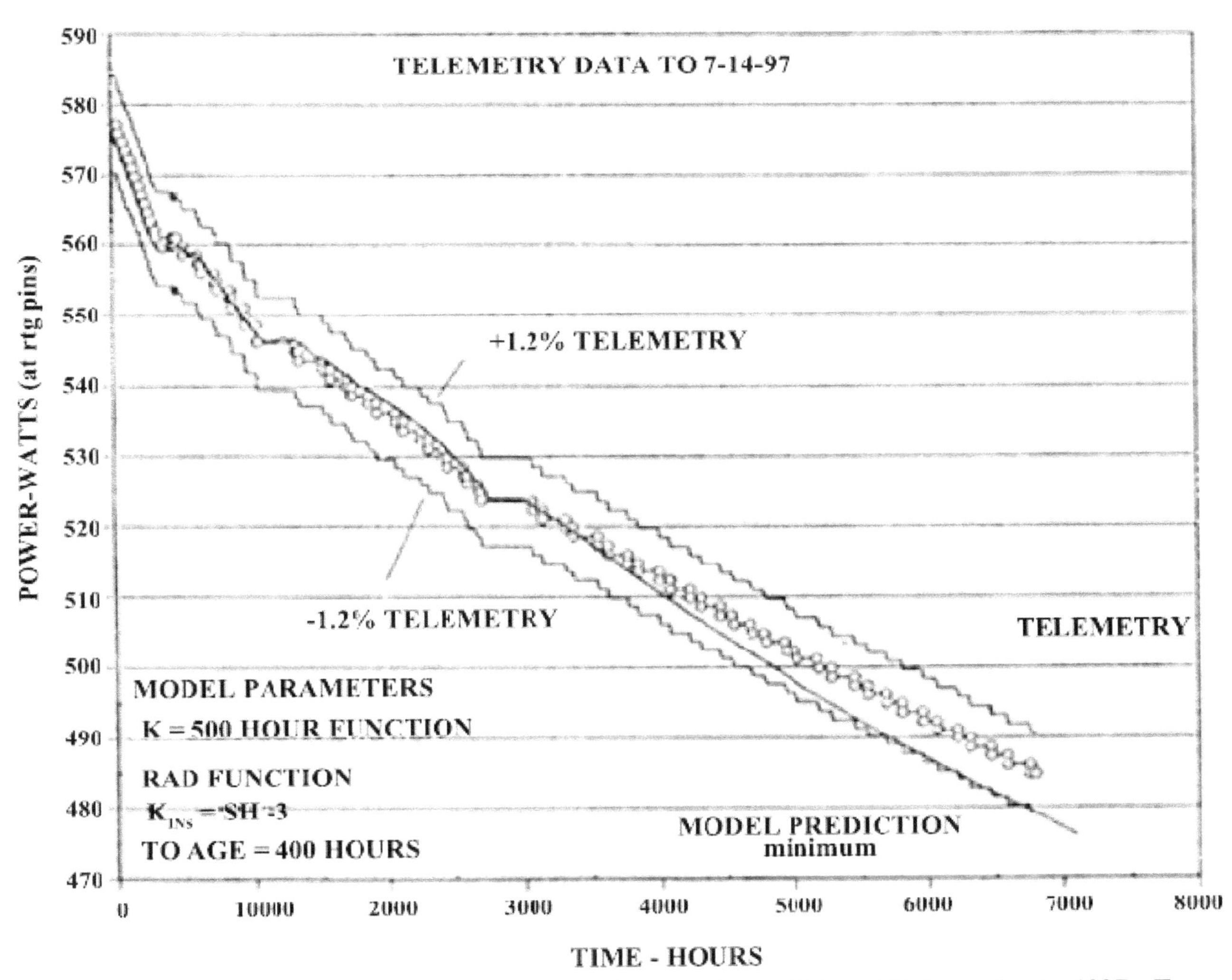

Fig. 9. Total Galileo GPHS-RTG power output (summation of F-1 and F-4) to 14 July 1997. ***From Bennett 2006.***

Ulysses, after a 4.5 year delay caused by the Challenger accident, also performed flawlessly. Fig. 10 shows the power output. Again, the excellent performance of the power system allowed the mission to be extended several times. Both BOM and EOM values exceed the requirements; initial BOM power of 284 W_e exceeded the requirement of 277 W_e and EOM was 248 W_e which exceed the 245 W_e requirement.

Cassini, with three GPHS-RTGs, performance is shown in Fig. 11. The BOM power of 887 W_e exceeded the specification requirement of 826 W_e. The projected power at 16 years after BOM is 640 W_e; this will exceed the specification requirement of 596 W_e.

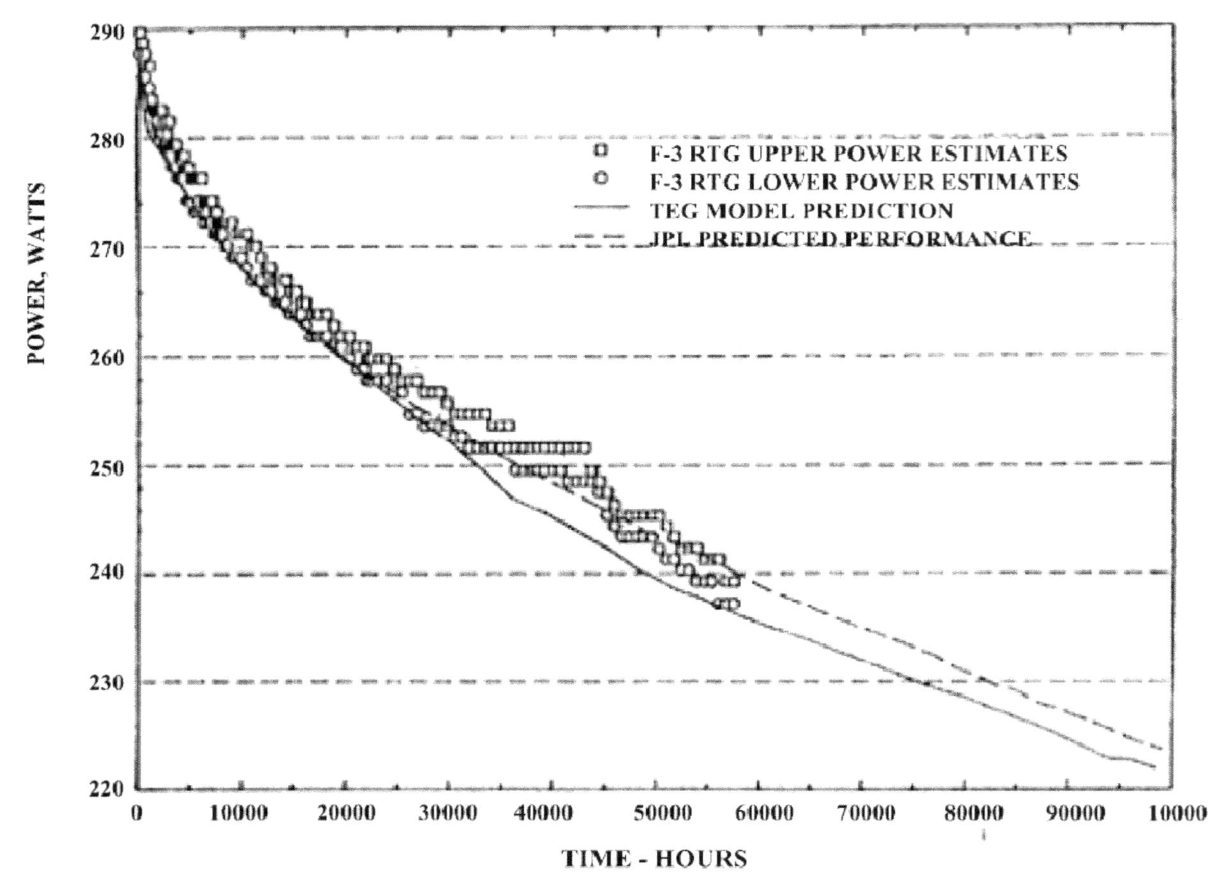

Fig. 10. Ulysses GPHS-RTG (F-3) power output to 1 April 1997.

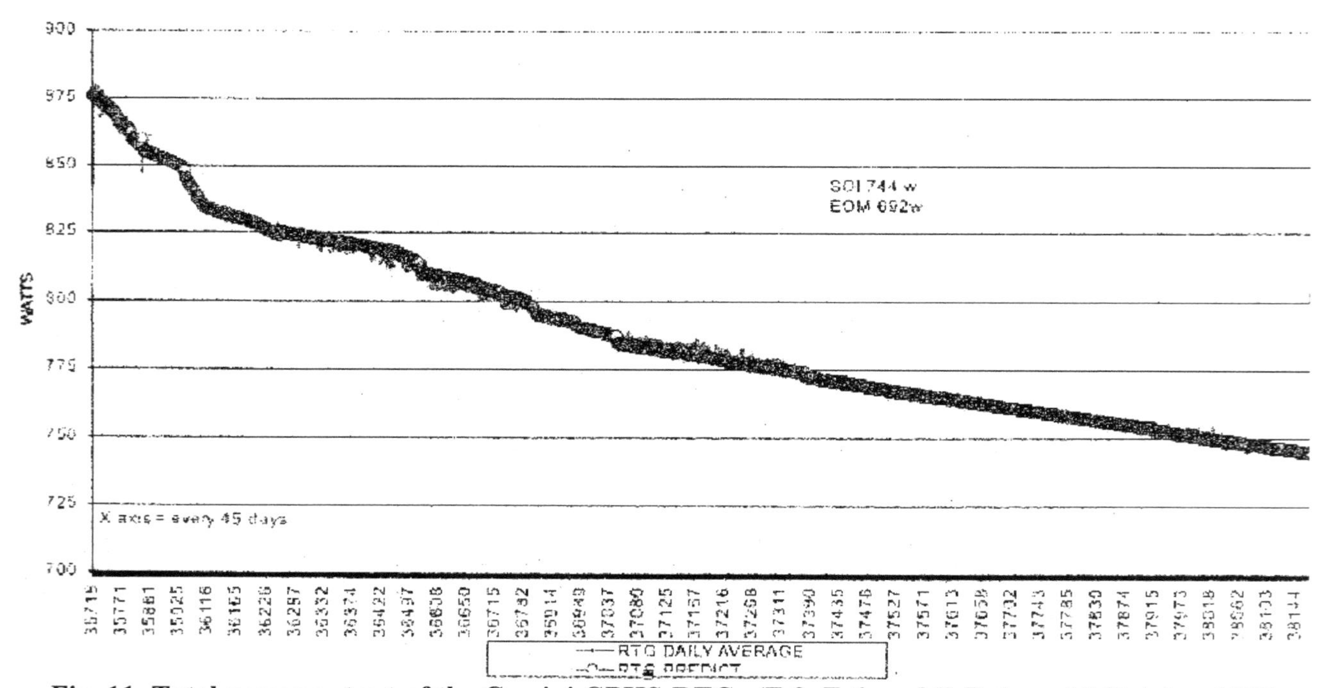

Fig. 11. Total power output of the Cassini GPHS-RTGs (F-2, F-6, and F-7) from 15 October 1997 to 30 June 2004. (Saturn Orbit Insertion, SOI, was 1 July 2004) ***Courtesy of Jet Propulsion Laboratory.***

GPHS-RTG Summary

The US has successfully used seven GPHS-RTGs in planetary missions. All of these systems have operated without incidents.

Safety related features are built into the design for normal, accident, and post-accident conditions. These include:[16]

- minimizing the release and dispersion of the PuO_2 fuel, especially of biologically significant small respirable particles;
- minimizing land, ocean and atmosphere contamination, particularly in populated areas; and,
- maximizing long-term immobilization of the PuO_2 fuel following postulated accidents.

Safety was accomplished by:

- Using a modular fuel approach in the design. The GPHS-RTG is composed of a group of eighteen fuel modules. Each module is designed to release the individual aeroshell modules in case of inadvertent reentry into the Earth's atmosphere. This minimizes the terminal velocity and the potential for fuel release on Earth impact. The converter uses an aluminum alloy to ensure melting and breakup of the converter if reentry occurs, resulting in release of the modules.

- The GPHS aeroshell module incorporates a three-dimensional carbon-carbon Fine Weave Pierced Fabric that was originally developed for reentry nose cones. The module and graphite components provide reentry and surface impact protection to the Iridium fueled clad in case of an accidental sub-orbital or orbital reentry. Later GPHS-RTGs have been modified to include additional graphite material between the GISs and strengthen the module under impact and reentry conditions.

- The PuO_2 fuel has a high melting temperature (2,673 K), very insoluble in water, and fractures into largely non-respirable chunks upon impact.

Multi-Mission Radioisotope Thermoelectric Generator (MMRTG)

NASA is planning to launch in 2011 a large, highly capable rover, the Mars Science Laboratory, and land it on the surface of Mars in 2012. This is part of the Mars Exploration Program to:

- Determine if life exists or has ever existed on Mars;
- determine if life exists today,
- determine if life existed on Mars in the past,
- assess the extent of organic chemical evolution on Mars.

- Understand the current state and evolution of the atmosphere, surface, and interior of Mars;
- characterize the current climate and climate processes of Mars,
- characterize the ancient climate of Mars,
- determine the geological processes that have resulted in formation of the Martian crust and surface,
- characterize the structure, dynamics, and history of the planet's interior.1

- Develop an understanding of Mars in support of possible future human exploration;
- acquire appropriate Martian environment data such as those required to characterize the radiation environment,
- conduct *in situ* engineering and science demonstrations.

Fig. 12 is an illustration of the rover showing the MMRTG location. The rover is deigned to operate for at least one Mars year (687 Earth days) over a wide range of candidate landing sites. Power is to be provided by the Multi-Mission Radioisotope Thermoelectric Generator (MMRTG).

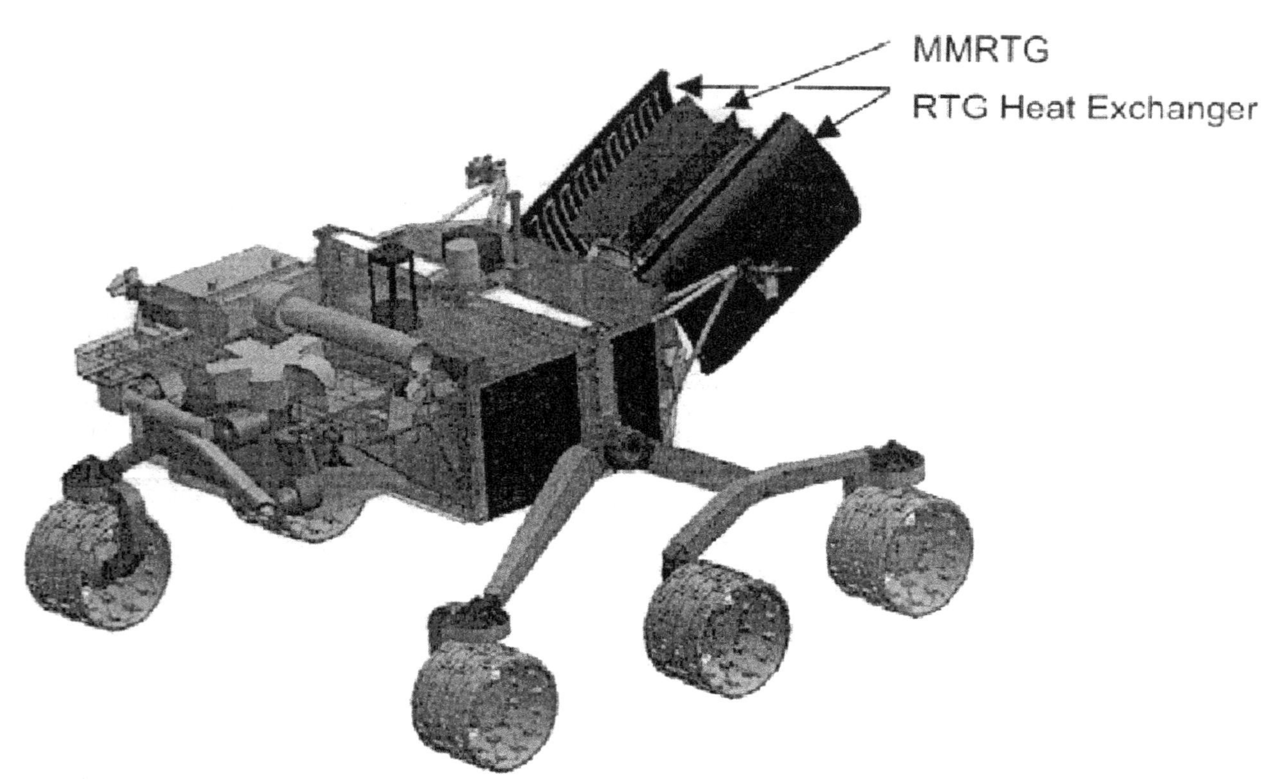

Fig. 12. Mars Science Laboratory with MMRTG power system. ***From Alan V. von Arx 2006.***

MMRTG Design

Table 7 list the design requirements for the MMRTG.[17] The requirements included the worst-case thermal environment of the Mars surface of 270 K. The multi-mission aspect included operating in an oxidizing atmosphere, space vacuum, and performing other missions beyond the Mars Science Laboratory.

Safety was an important factor in the MMRTG design. The MMRTG was designed to provide for containment of the PuO_2 fuel to the extent feasible during all mission phases, including ground handling, launch, and unplanned events such as reentry, impact, and post-impact situations. This included:

- minimizing the release and dispersion of the PuO_2 fuel, especially of biologically significant, small, respirable particles;

- minimize land, ocean and atmosphere contamination, particularly in populated areas; and

- maximize long-term immobilization of the PuO_2 fuel following postulated accidents.

This led to the following safety features built into the MMRTG design:

- The thermoelectric converter housing is made of aluminum alloy to ensure melting and breakup of the converter upon reentry. This results in release of the GPHS modules.

- The GPHS module and its graphite components are designed to provide reentry and surface impact protection to the iridium fueled clad in case of accidental sub-orbital or orbital reentry. The aeroshell and GIS are composed of a three-dimensional carbon-carbon Fine Weave Pierced Fabric, developed originally for reentry nose cone material.

- The iridium clad material is chemically compatible with the graphite components of the GPHS module and the PuO_2 fuel over the operating temperature range of the MMRTG. Iridium is used because of its high-temperature (2,727 K) melting level and excellent impact response.

- PuO_2 fuel has a high melting temperature (2,623 K), is very insoluble in water, and fractures into largely non-respirable chunks upon impact.

Table 7. Key requirements for system selection. ***From Alan V. von Arx 2006.***

Requirement	Value	Notes - Derived Requirements
No. GPHS Modules	8	- Use 8-GPHS modules
BOM Power (W_e)	> 110	- BOM defined as just after launch
		- No benefit for power greater than 110 W
Environment	Multimission	- Use a sealed unit (SNAP 19 derived)
		- Size for worst case thermal environment (270 K Mart day)
Power Degradation	Minimize	- Target less than 25% power loss in 14 yrs
Mass	<45 kg	
Volume	As small as possible	
Specific Power	Maximize	- Target greater than 3 W/kg @ BOM
Lifetime (yrs)	> 14	- Use approach with proven lifetime capability (SNAP derived)
Integration/constraints	Define	- Define thermal/interface requirements for mission ph
Reliability	Maximize	- No credible single point failures, single fault tolerant circuit design
Safety	GPHS release	- Verify materials of construction will melt away

The MMRTG, the latest generation of RTG, is illustrated in Fig. 13. It is designed to use a heat source composed of eight General Purpose Heat Source (GPHS) modules. This will provide approximately 2,000 watts of thermal power and 120 watts of electrical power. It contains a total of 4.8 kg of plutonium dioxide.

The thermoelectric conversion system uses designs from the two Viking spacecraft that landed on Mars in 1976. These are the ones described under SNAP-19. The thermocouple materials are lead telluride (PbTe) and telluride-antimony-germanium-silver (PbTe-TAGS). An illustration of the thermoelectric couple is seen in Fig 14. The transition temperature of 700 K was optimized to minimize degradation while still meeting the BOM power requirement with margin. Resulting power loss over the 14-year mission life is less than 25%. Sixteen percent of the power loss over a fourteen-year mission is due to natural fuel decay and 8.5% due to thermoelectric degradation. A total of 16 thermoelectric modules are used, with eight rows of six couples each. For fault tolerance, a two-string series-parallel arrangement of the thermoelectric couples, cross-strapped at each couple is used. A trade-off between the radiator size/mass and life of the thermoelectrics resulted in the selection of a cold junction temperature of 583 K.

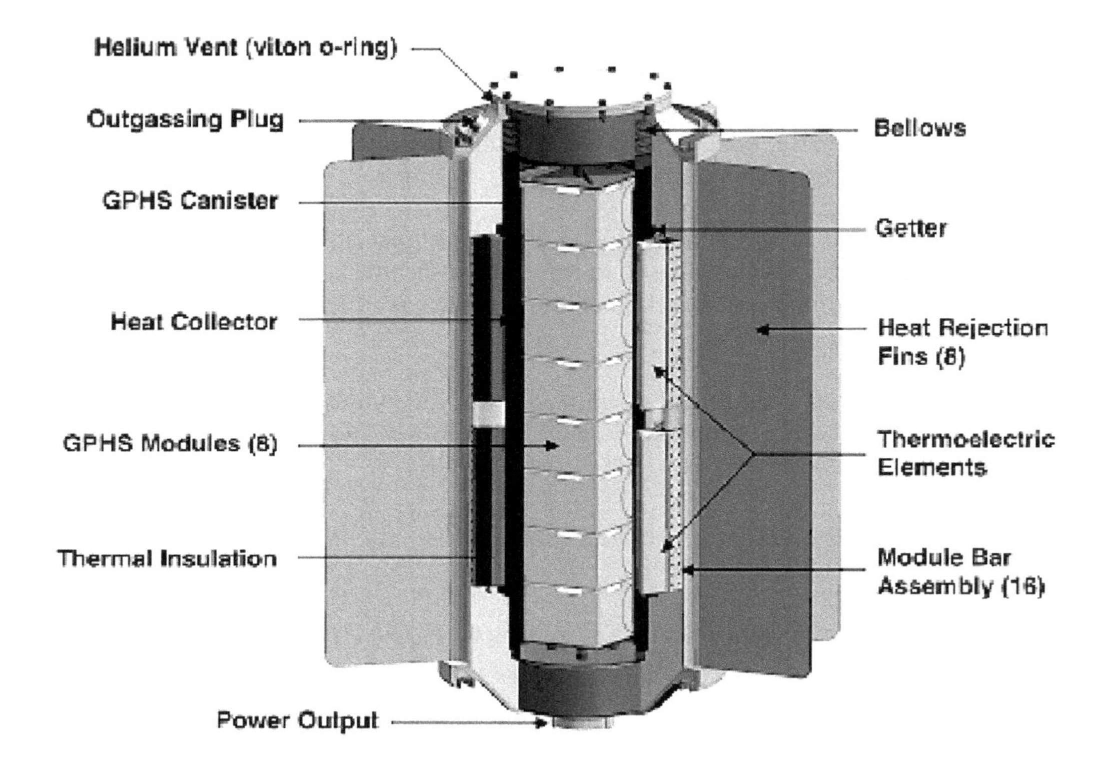

Fig. 13. Illustration of MMRTG design. ***Courtesy of Department of Energy***

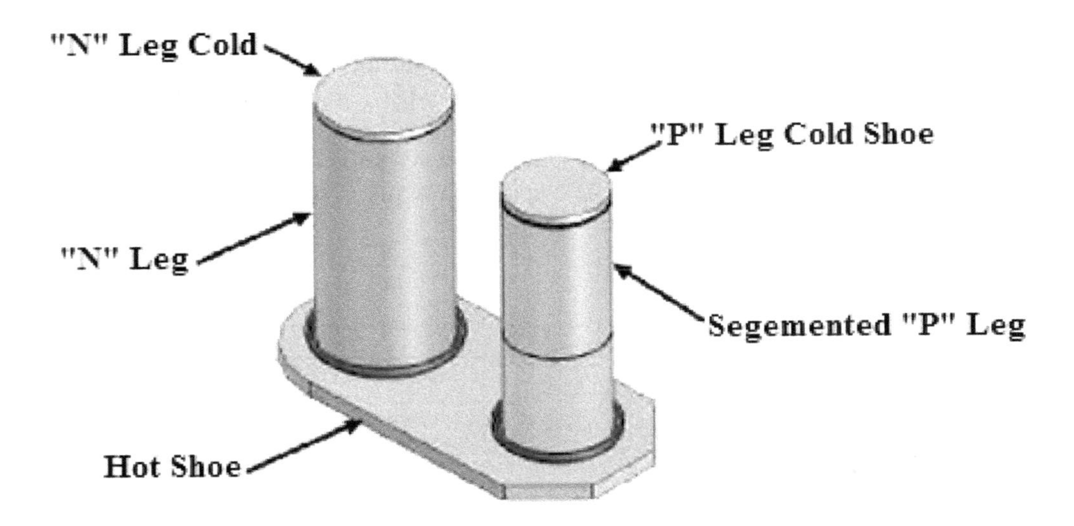

Fig. 14. MMRTG thermoelectric couple. ***Source Department of Energy.***

The geometry of the heat rejection system is illustrated in Fig. 15. Eight conductive fins located radially around the converter housing are used in the heat rejection system. Aluminum is the basic construction material. Al 2219 was selected for strength while the fins are made from Al 6061 for the higher thermal conductivity. The Al-6061 fin are approximately 20.3 cm in length along the full height axis of the converter and tapered to provide the lowest mass and volume with a root thickness of 0.41 cm.

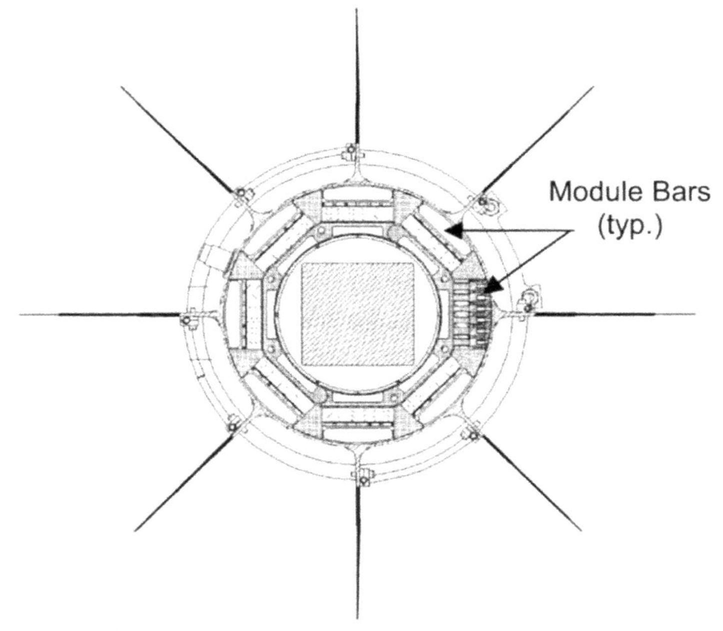

Fig. 15. Radial cross sectional view showing modular bars directly inside the fins. ***From Alan V. von Arx 2006.***

The heat source support design utilizes the heritage SNAP-19 approach of pre-loaded Min-K to constrain the GPHS modules. Design features include: Min-K provides a low conductance path to minimize parasitic losses; the heat source preload is maintained completely within the sealed cavity, with no screw heads penetrating the enclosure; Min-K acts as a compliant spring to store preload potential energy over the mission length and mitigate shock transmitted to the GPHS modules; and Min-K has demonstrated compatibility with heat source and converter material over 40 years.

The MMRTG requires some supplemental cooling inside the vehicle/spacecraft (see Fig. 16). The aluminum tubes attached near the fin roots will be used for heat removal during fueling, sterilization, pre-launch and cruise integration. While integrating the MMRTG into the MSL, it will be necessary to maintain the temperatures of the nearby electronics below their limits. This involves the addition of a Freon coolant loop. The Freon coolant would have flashed when it contacted the hot MMRTG and heated up the electronics once circulation is started. This led to a "double tube" approach with a second cooling loop to protect the electronics during startup.

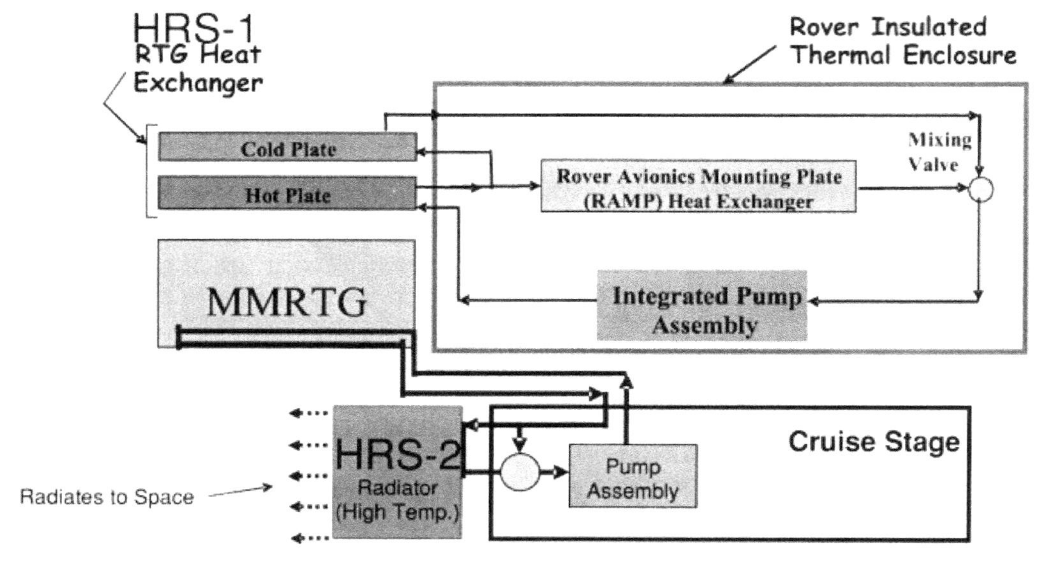

Fig. 16. MSL thermal control. ***Alan V. von Arx, 2006.***

On the Mars surface operations, another Freon pumped loop is employed. An RTG Heat Exchanger is configured to pick up heat from the MMRTG and transfer it to the Rover electronics in a cold environment for convection and radiation to the environment.

Safety considerations in the design includes dispersal of the GPHS stacks in case of an inadvertent reentry. The aluminum housing is design to melt due to the dissipation heating,. The GPHS stack will disassemble and each module will impact the earth as an individual entity.

Design results are given in Table 8. The MMRTG generator is about 64 cm in diameter by 66 cm long and weighs about 43 kg.

Table 8. MMRTG design parameters. ***From Alan V. von Arx 2006.***

Parameter	Units	Value	Notes
Hot Junction Temperature	°C (°F)	541 (1,005)	Within range of current database.
TAGS Transition Temperature	°C (°F)	426 (800)	TAGS is susceptible to degradation at high temperatures; optimized at this value.
Cold Junction Temperature	°C (°F)	210 (410)	Optimized between performance and heat rejection system mass.
Average Fin Root Temperature	°C (°F)	191 (375)	Average temperature of the housing in contact with the module bars.
Number of couples		768	16 modules with 8 rows of 6 couples per module.
Number of parallel strings		2	A two-string series-parallel arrangement of the couples is used for high reliability. Cross-strapping at each couple.
Power Output, BOM	watts	126	
Power Degradation	%	24.8	Part due to radioactive decay; the rest primarily due to TAGS temperature.

MMRTG Performance

The MMRTG combines two well established subsystems. The heat source, GPHS, has had flight experience since 1989 on Galileo, Ulysses, Cassini and New Horizons planetary spacecraft and continued their outstanding performance after nineteen years operation. The thermoelectric converters, lead telluride and telluride-antimony-germanium-silver (PbTe-TAGS), were used on the SNAP-19 missions that included Viking 1 and 2 on the Mars surface and Pioneer 11 planetary spacecraft. Viking operated successfully for up to 6 years and Pioneer for twenty-two years. In addition, the balance-of-plant components uses well established materials and technologies.

Fig. 17 is an engineering unit for validating the readiness of MMRTG for flight. Its purpose was to validate the MMRTG design and its readiness for the Mars Science Laboratory mission.

Power performance is relatively insensitive to Martian diurnal cycle where daily temperatures range from 170 to 270 K. The Martian day/night cycle indicates less than a 3 W difference. For Mars missions, the lower temperatures experienced during Martian night will reduce degradation, enhancing EOM power output.

Environmental Impact Evaluation[18]

The nuclear risk assessment for the MSL mission considered: (1) potential accidents associated with the launch, and their probabilities and accidental environments; (2) the response of the MMRTG to such accidents in terms of the amount of radioactive materials released and their probabilities; and (3) the radiological consequences and mission risks associated with such released. The risk assessment was based on a typical MMRTG radioactive material inventory of about 58,700 Ci of primarily plutonium-238. The categories used to assess risk are either unlikely, very unlikely or extremely unlikely. These are defined as:

- unlikely: 1 in 100 to 1 in 10 thousand;
- very unlikely: 1 in 10 thousand to 1 in 1 million; and
- extremely unlikely: less than 1 in 1 million.

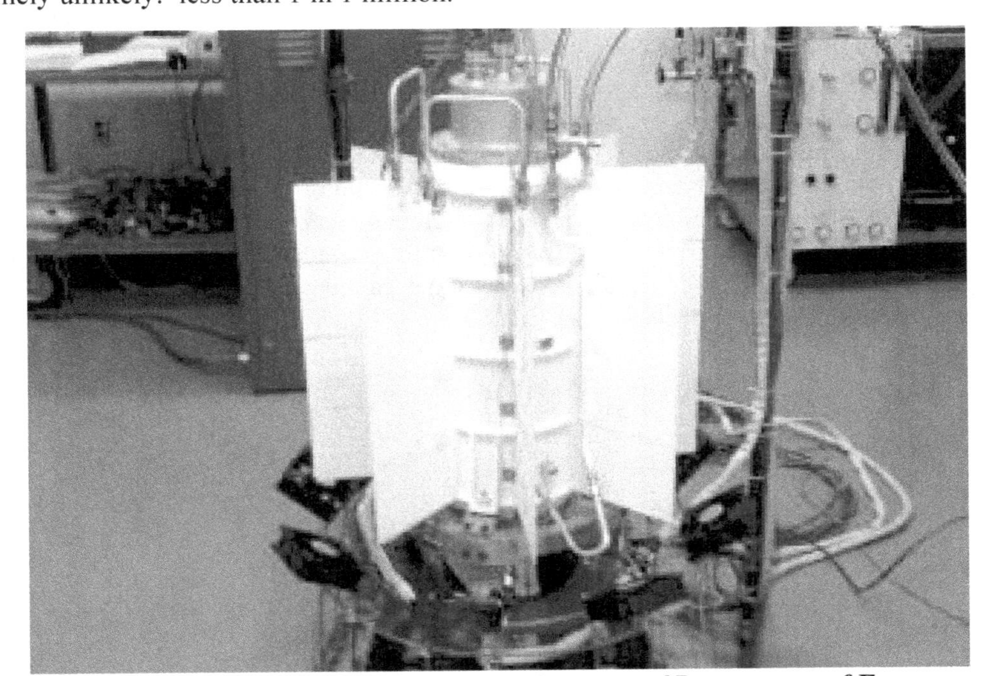

Fig. 17. MMRTG Engineering Unit. ***Courtesy of Department of Energy.***

The radiological consequences of a given accident that results in a release of radioactive material were calculated in terms of radiation doses, potential health effects, and area contaminated at or above specified levels.

- Phase 0 (Pre-Launch) and Phase 1 (Early Launch): A launch related accident during these periods could result in ground impact in the launch area with some release of PuO_2 fuel from the MMRTG. Prior to launch, the most likely result of a launch vehicle problem would be a safe hold or termination of the launch. After launch, the most significant launch vehicle problems would lead to the automatic or commanded activation of on-board safety systems, resulting in destruction of the launch vehicle. These Phases have a total probability of an accident resulting in a release of about 1 in 420--placing them in the unlikely category. The maximum dose received by an individual within the exposed population would have a mean value of about 0.14 rem. This is equivalent to about 40 percent of the normal annual background dose received by each individual in the U.S. during a year. The mean collective dose that would be received by all individuals with the potentially exposed local and global populations would be about 740 person-rem--or about 0.4 additional latent cancer fatalities within the entire group of potentially exposed individuals. About 60% of the risk to the exposed population is local (within 100 km of the launch).

- Phase 2 (Late Launch): A launch accident during this period would lead to impact of debris in the Atlantic Ocean with no release of PuO_2 fuel. Undamaged GPHS modules would survive water impact at terminal velocity. There would be no health consequences.

- Phase 3 (PreOrbit/Orbit): The total probability of an accident resulting in a release is about 1 in 1,100--considered an unlikely event. The maximum mean value dose received by an individual close to the impact site would be about 0.23 rem. This is equivalent to about two-thirds of the normal annual background dose received by each person in the U.S. during a year. The collective dose received by all individuals within the potentially exposed global population would be about 6 person-rem. This is about 0.003 additional latent cancer fatalities within the exposed population.

- Phase 4 (Orbit/Escape) A launch accident after attaining parking orbit could result in orbital decay re-entries from minutes to years after the accident. This is considered an unlikely event, about 1 in 830. The maximum mean value dose received by an individual to the impact site would be about 0.7 rem. This is about twice normal background. The collective dose received by all individuals within the potentially exposed global population is about 64 person-rem or 0.03 additional latent cancer fatalities within the exposed population.

The overall mission accident probability associated with the MSL mission is the sum of the accident probabilities for several accident conditions for all mission phases (see Table 9). Across the entire mission, the accident results are about 1 in 220--category unlikely. The maximum dose received by an individual within the potentially exposed population would vary with accident and meteorological conditions. The mean value is about 0.3 rem, or about 80% of the normal U.S. background dose received annually by each member of the population. The collective dose received by all individuals within the potentially exposed population would be about 400 person-rem. This is statistically an additional 0.2 latent cancer fatality among the exposed population.

Table 9 Summary of health effect mission risks. ***Source Final Environmental Impact Statement for the Mars Science Laboratory Mission.***

Mission Phase[(a)(b)]	Accident Probability	Conditional Probability of a Release	Total Probability of a Release	Mean Health Effects	Mission Risks
0: Pre-Launch	7.6×10^{-4}	0.78	Very Unlikely (5.9×10^{-5})	0.21	1.3×10^{-4}
1: Early Launch	4.4×10^{-3}	0.54	Unlikely (2.4×10^{-3})	0.37	8.7×10^{-4}
2: Late Launch	1.3×10^{-2}	-	-	-	-
3: Pre-Orbit/Orbit	1.1×10^{-2}	0.086	Unlikely (9.0×10^{-4})	0.0032	2.9×10^{-4}
4: Orbit/Escape	4.8×10^{-3}	0.25	Unlikely (1.2×10^{-3})	0.032	3.8×10^{-5}
Overall Mission	3.3×10^{-2}	0.14	Unlikely (4.5×10^{-3})	0.2	9.1×10^{-4}

Source: DOE 2006a

a. The table presents composite of the results for the Atlas V541 and the Delta IV Heavy determined by taking the probability-weighted value of the two sets of results, treating the conditional probability of having a given launch vehicle as 0.5. Accident probabilities are the average of individual values for the two vehicles. Based on the current state of knowledge, the specific accident probabilities for the accident conditions for each vehicle are expected to be similar.

b. The reported values are within a factor of two of the high-end values when the results for each launch vehicle are considered separately.

c. Notes: Differences in multiplications and summations are due to rounding of results as reported in DOE 2006a. Probability categories, i.e. unlikely, very unlikely, defined by NASA.

Summary

Use of seven GPHS-RTGs (two units on Galileo, one unit on Ulysses, three units on Cassini and one unit on New Horizons spacecraft) have proven them to be reliable and predictable power sources. In the Galileo and Ulysses missions, the missions were extended several times as a result of the successful operation of the power system. The Galileo mission ended with the insertion of the probe into the atmosphere of Jupiter. Ulysses, Cassini and New Horizons missions, based on the performance to date, should meet or exceed all the remaining power performance requirements despite their harsh operational environments.

The MMRTG is on track for launch in 2011 as part of the Mars Science Laboratory. It is designed to operate in the highly oxidized Martian environment. As such, it will use the proven GPHS power source and power conversion subsystem so successful in the Viking 1 and 2 missions.

Chapter 5

Alternate Power Conversion Subsystems

The major limitation in the use of radioisotope power systems is the high cost and availability of plutonium-238. Higher efficient and lighter weight converter subsystems would reduces the quantity of plutonium-238 required, radioisotope generator cost, and the launch cost associated with the power system. In addition, more efficient electrical converters can extend the operational electric power range of radioisotope systems from about one kilowatt-electric to ten kilowatts-electric. Various programs, past and present, have had as their goals the development of higher efficient conversion subsystems.

Radioisotope heat sources can be combined with dynamic thermal-to-electrical energy conversion devices to achieve efficiencies about three times higher than those obtainable with thermoelectric converters.[1, 2, 3] This is a significant advantage because fuel costs are a major factor in the overall costs of radioisotope power supplies. Dynamic converters take several forms--all of which have their advantages and disadvantages. In particular, Brayton, Rankine and Stirling cycles have gone through development cycles for use as space power electrical generators. The principles of these generators will be covered and their development programs summarized.

NASA's Lewis Research Center initiated the development of a 2 to 10 kW_e Closed Brayton Cycle power system in 1965. In the 1975 - 1980 time period, the Department of Energy sponsored programs that included ground demonstrations units of both Brayton and Rankine technologies. The basic technologies of both cycles were demonstrated. Results included a ground demonstration of a Brayton cycle engineering unit that operated for over 1,000 hours; no problems were identified. In the parallel program on Rankine cycles, a ground-based system demonstrated that the use of an organic working fluid, such as Dowtherm A or toluene, could provide power in the 1 to 10 kilowatt-electric range--these organic fluids avoided the materials and corrosion problems of liquid-metal systems. As part of the program, the ground demonstration configuration of an organic fluid Rankine cycle radioisotope system completed a 2,002 hour endurance test with no component failures and also operated for 11,146 hours of overall testing.[4] Efficiencies in the 20 to 25% range were demonstrated in both conversion programs. Depending upon the design power levels, specific powers of 6.6 W_e / kg to 8.8 W_e / kg are projected.[5]

Also studied for advanced radioisotope generators are the Alkali-Metal Thermal-To-Electric Converters (AMTEC) in the AMTEC Radioisotope Power System (ARPS) program. The technology readiness level for AMTEC is rated as 3 (analytical and experimental critical function and/or characteristic proof-of-concept).

Currently, under development is the 110 W_e Advanced Stirling Radioisotope Generator (ASRG). The ASRG project has proceeded through a systems requirements review in May 2009. A final design review is anticipated in 2010, with flight systems being available for a mission in late 2013.[6] Also, an Advanced Thermoelectric Converter program to demonstrate a 35-65% improvement in thermoelectric converter efficiency over the GPHS-RTG and MMRTG has a near-term goal to demonstrate an 18 thermocouple unit.

Principles of Dynamic Thermal-to-Electric Conversion Cycles[7]

A heat engine can be defined as a controlled-mass, thermodynamic device that receives thermal energy from a source, converts a portion of this energy into mechanical work (which is transferred across the system boundaries to the surrounding application), and then rejects the unused amount of input thermal energy to a lower temperature sink. The output work of interest here is in the form of electrical power. The second law of thermodynamics establishes the maximum amount of input thermal energy that can actually be converted into work. Real heat engines fall short of this performance optimum as a result of irreversibilities caused by the heat transfer process with large temperature differentials and working fluid-induced frictional effects. The working fluid within the heat engine undergoes cyclic operation; after some period of time all the working fluid within the device returns to its initial state. A generic heat engine that operates between a high temperature heat source (T_H) and a low temperature heat sink (T_L) is shown in Fig 1[8, 9]. This type of engine is called a 2*T* heat engine, since thermal energy is transferred into the system at one temperature and rejected from the system at another (lower) temperature.

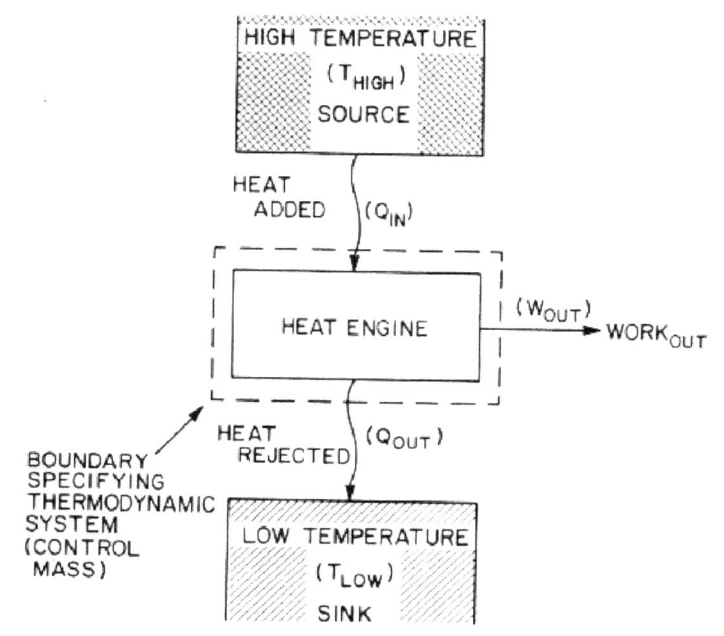

Fig. 1. The reversible two temperature heat engine.

Efficiency expresses the effectiveness of using the heat available. The efficiency of a system can be defined as:

$$\eta \equiv \frac{\textbf{some useful effect}}{\textbf{energy that must be expended to achieve that effect}} \tag{1}$$

The maximum thermal efficiency of a heat engine is called its Carnot efficiency (η_{Carnot} or η_{th}). The Carnot cycle is a fundamental theoretical concept in dynamic energy conversion, because the thermal efficiency of this cycle is the maximum possible value for any heat engine operating between the same two temperature limits.[10, 11] Carnot (1796-1832) was a French military engineer who performed pioneering examinations of heat engine efficiencies in the early 19th century. Fig. 2 shows the pressure-volume and temperature-entropy diagrams of the Carnot cycle. In this idealized cycle, a working fluid experiences: reversible, isothermal heat addition from state I to state 2; reversible, adiabatic (i.e., isentropic) expansion from 2 to 3 during which work is transferred from the system to its environment; reversible, isothermal heat rejection from state 3 to state 4; and finally, a reversible, adiabatic (i.e., isentropic) compression process from 4 to I during which work is transferred to the system from the surroundings. By definition, an isentropic or constant entropy process is a reversible, adiabatic process

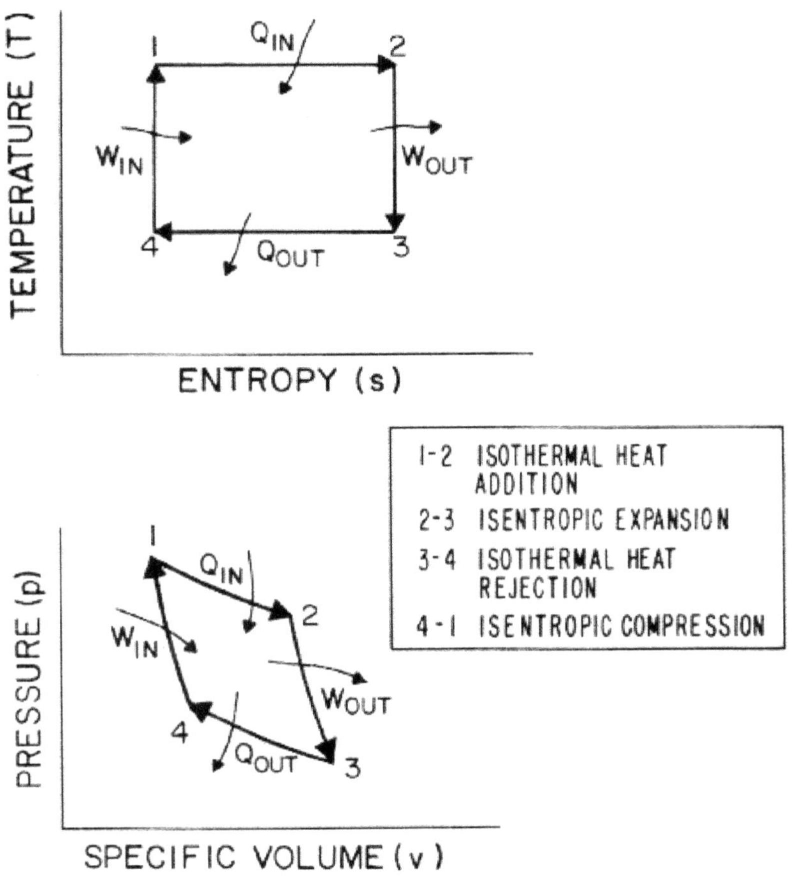

Fig. 2. Carnot cycle.

In theory, isothermal heat addition and rejection can be achieved by using suitable heat exchangers, while isentropic turbines and compressors or pumps may be used to transfer energy as work to or from the working fluid. The theoretical efficiency of this cycle is given by

$$\eta_{CARNOT} = \frac{WORK_{out}}{HEAT_{IN}} = \frac{T_H - T_L}{T_H} = 1 - T_L/T_H \quad (2)$$

where T_H is the higher absolute temperature (K) at which heat is added
T_L is the lower absolute temperature (K) at which heat is rejected.

Eq. 2, shows that the thermal efficiency of the system is greater, the higher the heat addition temperature (T_H); and the thermal efficiency is greater, the lower the heat rejection temperature (T_L). The Carnot efficiency represents the maximum thermal efficiency (h_{th}) of any heat engine operating between the same source (T_H) and the sink (T_L) temperatures.

Sometimes in treating a spacecraft's energy conversion system, thermal efficiencies (i.e., Carnot limitations) are separated from other system losses. These other losses are both electrical or mechanical. In that case, a power conversion system (PCS) efficiency can be defined as

$$\eta_{PCS} = \eta_{th} \cdot \eta_D \quad (3)$$

where η_{th} is the thermal efficiency
η_D is the device efficiency.

Brayton Isotope Power System Developments

Brayton Cycle Principles

The Brayton cycle is probably the more straight-forward of the dynamic conversion cycles. The working fluid is always in the gaseous state. A closed cycle gas turbine power system is illustrated in Fig. 3. An idealized Brayton cycle has heat addition and rejection occurring at constant pressure, while expansion and compression processes are assumed to be isentropic. The working fluid process is: constant pressure heat addition from state 1 to 2; isentropic expansion in the turbine from state 2 to 3 (a portion of the output work goes to compressing the working fluid); constant pressure heat rejection from state 3 to 4; an isentropic compression from state 4 to 1.

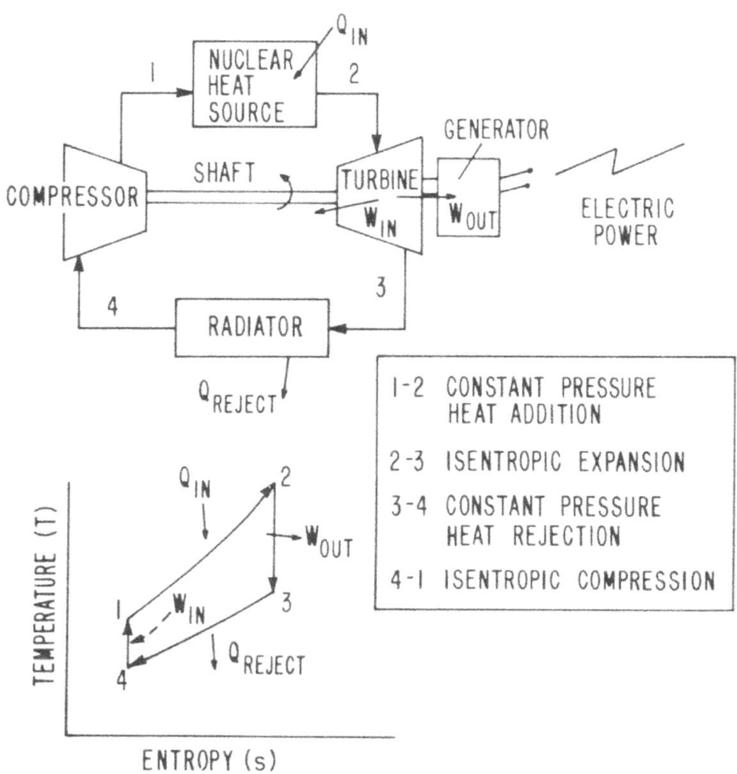

Fig. 3. Closed Brayton cycle (ideal).

The thermal efficiency of the ideal closed Brayton cycle is given by: (neglecting working fluid potential and kinetic energy changes)

$$\eta_{th} = \frac{(W_{out})_{turbine} - (W_{in})_{compressor}}{Q_{in}} \tag{4}$$

or,

$$\eta_{th} = \frac{(h_2 - h_3) - (h_1 - h_4)}{(h_2 - h_1)} \, . \tag{5}$$

If the working fluid is assumed to be an ideal gas, the thermal efficiency is given by

$$\eta_{th} = 1 - (p_1/p_4)^{(1-k)/k} \tag{6}$$

where $k = c_p / c_v$, the ratio of the specific heats of the gas at constant pressure and volume respectively.

To improve the thermal efficiency of the basic closed Brayton cycle, a regenerator (or recuperator) can be incorporated into the cycle, see in Fig. 4. In the regenerator, the hot gases, which are exhausted from the turbine (i.e., state 3), preheat the working fluid as it exits the compressor (state 6) and returns to the nuclear power supply (state 1).

Another way to improve Brayton cycle efficiency is through the use of multistage compression with intercooling and multistage expansion with reheat. In an ideal intercooler (see Fig. 5) the gas is cooled to the initial inlet temperature T_1 before entering the next state; that is, $T_1 = T_3$ and $p_2 = p_3$.

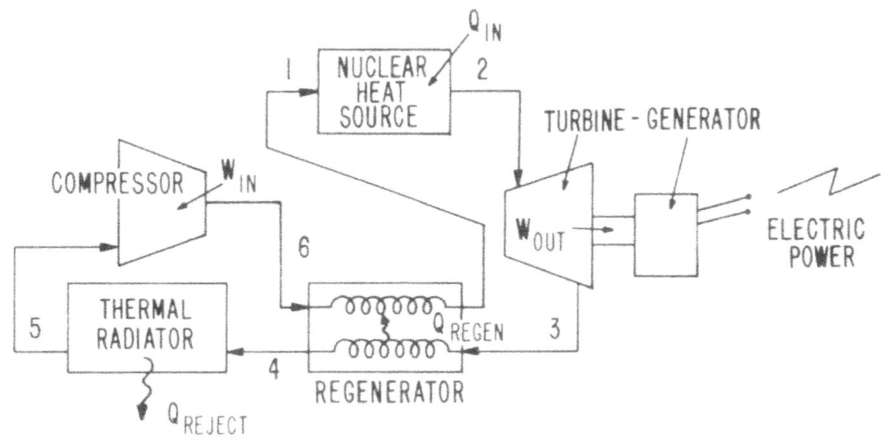

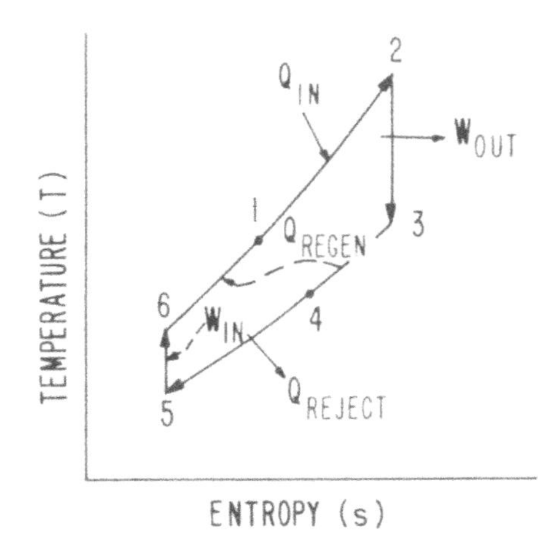

Fig. 4. Closed Brayton cycle with regeneration (ideal).

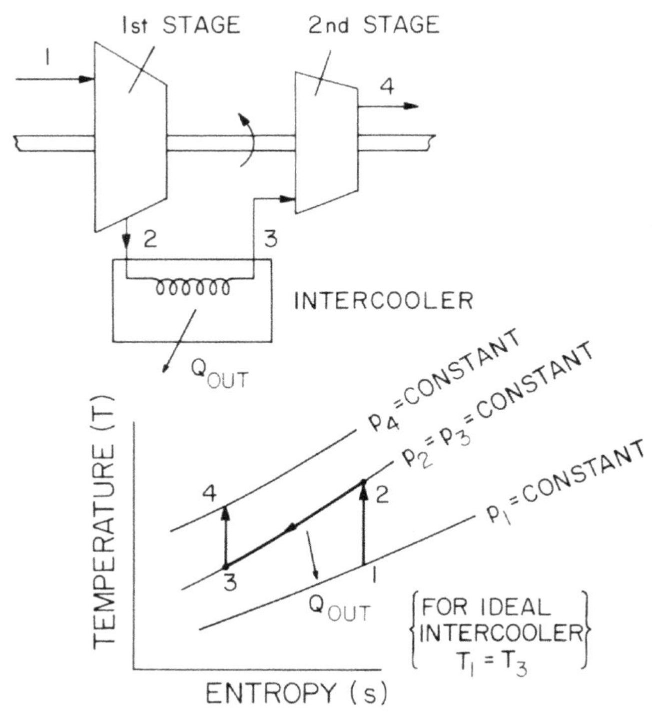

Fig. 5. Intercooling: ideal two-stage compressor.

The combination of reheating, intercooling and regeneration provides the best practical Brayton cycle efficiency. If regeneration is used along with a very large number of reheating stages and intercooling states, then, in principle, all thermal energy input eventually appears in the reheat-heat exchangers (when the gas is at its maximum temperature) and all heat rejection occurs in the intercoolers (when the gas is at its lowest temperature). This condition can approach the Carnot efficiency of a reversible $2T$ heat engine (see Fig. 6).

Brayton Isotope Power System (BIPS)

The Brayton Rotating Unit (BRU) Project (1968 to 1978) was aimed at a high efficiency power conversion system for isotope, reactor, and solar receiver heat sources. The design was for power levels between 2.25 to 10.5 kW_e. The project successfully demonstrated efficiency up to 32%, operated at turbine inlet temperature to 1,144 K, compressor inlet temperature of 300 K, and maximum pressure of 310 kPa. In the 1974 to 1978 time period, a smaller version of the BRU unit, called mni-BRU, was developed to demonstrate efficiency up to 30% in power levels from 500 to 2,100 W.[12]

The Mini-BRU system formed the basis of the Brayton Isotope Power Systems (BIPS). It had the objective to demonstrate a system at 1.3 kW_e with a mass of 204 kg. The major components of the system are shown in Fig. 7.[13] The system was designed for a turbine inlet temperature of 1,144 K, compressor inlet temperature of 300 K, and maximum pressure of 738 kPa. The higher pressure resulted in a smaller rotating assembly and higher shaft speed (52,000 rpm).

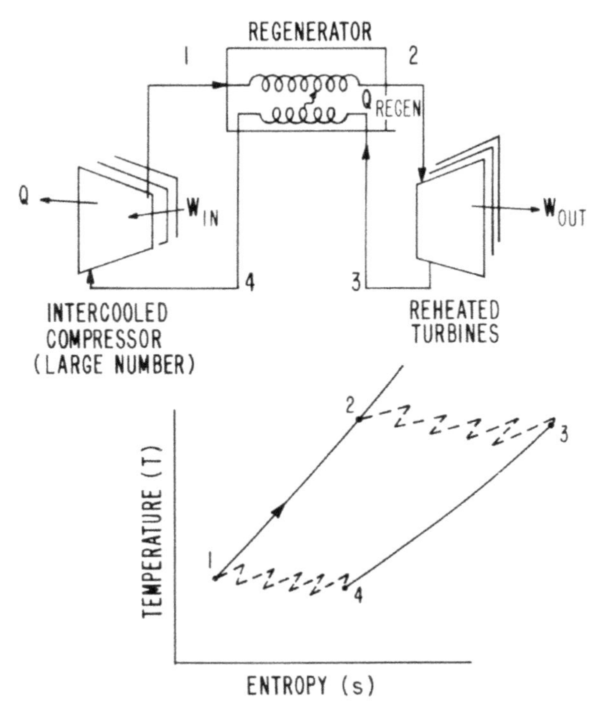

Fig. 6. Regenerative Brayton cycle (ideal) with a large number of reheating and intercooling stages.

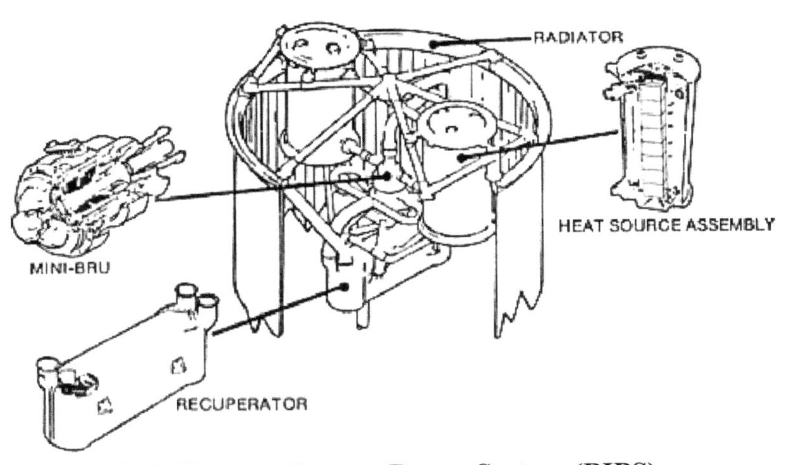

Fig.7. Brayton Isotope Power System (BIPS).

The heart of the converter, the Brayton Rotating Unit, is a compressor-turboalternator unit. (Fig. 8). The turboalternator-compressor is mounted on a single rotating shaft. This high-speed unit is the only rotating component in the power converter unit. These are single-stage radial compressors and turbines. The shaft is supported on gas bearings and the alternator is brushless, with no metal-to-metal contact during operation. The gas bearings are self-actuated, compliant foils (see Fig. 9)[14]. When the mini-BRU is started the foils remain in contact with the moving surface until the boundary layer forms and the journal becomes airborne. Cooling flow is provided to the bearings from compressor discharge gas bleed. The alternator can use compressor gas bleed or a separate liquid loop.

Work is extracted from the high temperature gas as it adiabatically expands in the turbine. Part of this work is used to run the compressor. The ratio of compressor input work $(W_{comp})_{in}$ to turbine output work $(W_{turb})_{out}$ is called the back work ratio of the Brayton cycle. That is,

$$\textbf{Basic Work Ratio} \equiv \frac{(\textbf{Compressor Work})_{in}}{(\textbf{Turbine Work})_{out}} \tag{7}$$

For the ideal Brayton cycle, this ratio becomes

$$\textbf{Back Work Ratio} = \frac{(h_1 - h_4)}{(h_2 - h_3)} \tag{8}$$

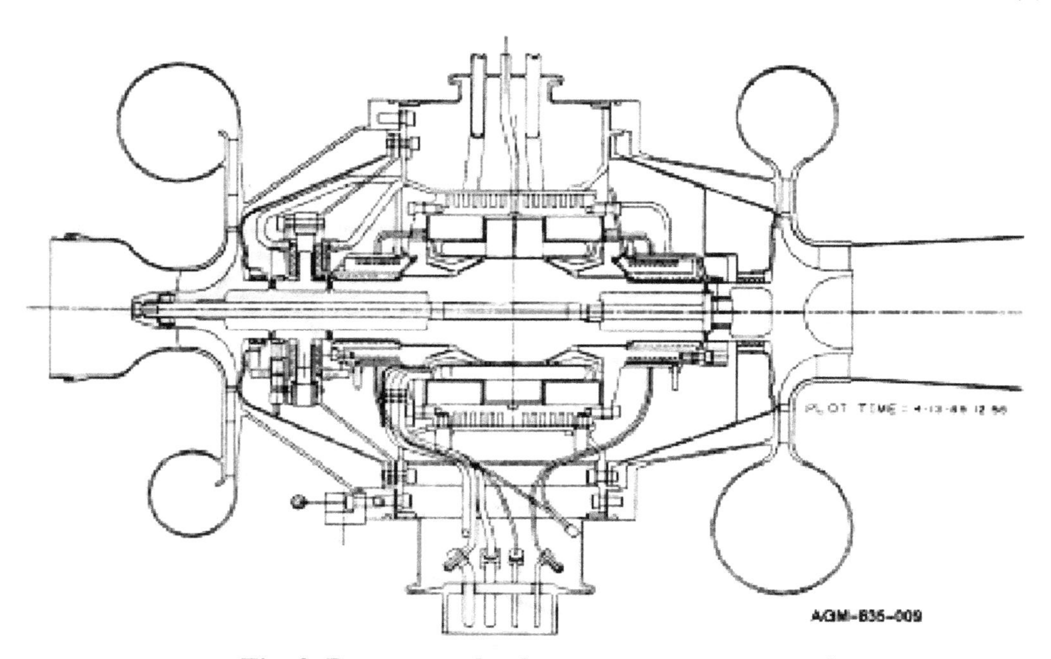

Fig. 8. Brayton turboalternator-compressor unit.

The alternator is a Rice brushless, non-rotating coil synchronous unit. The stator has a conventional three-phrase winding using a solid rotor. This reduces windage losses. Separating the magnetic and non-magnetic materials achieves structural strength, pole separation, windage loss reduction and electromagnetic damping. The field excitation coil is stationary and located in the frame of the machine. Fully redundant field coils are provided to meet high reliability requirements.

The working fluid was an inert gas to eliminate corrosion and corrosion by-products. A mixture of helium and xenon gasses with an effective molecular weight of 83.8 was found to optimize system mass and provide superior heat transfer properties over pure monatomic gases. The mini-BRU rotating speed was 867 Hz (52,000 rpm) and the turbine inlet temperature was 1,030 K. The radiator temperature was about 330 K. During testing, the compressor and turbine achieved efficiencies of 0.77 and 0.83, respectively.

The recuperator is of the plate-fin design. The assembly consists of stacked brazed panels of alternating low and high pressure counter-flow panels. Extensive accelerated life test of two hundred thermal shock cycles, equivalent to thousands of actual start cycles, were run with no appreciable increase in measurable internal or external leakage. The recuperator demonstrated an effectiveness of 0.98.[15]

During the BIPS program, a ground-test workhorse loop was assembled and tested. This loop completed over 1,000 h of testing with no system problems. To simulate space conditions, testing was performed in a vacuum chamber at approximately 10^{-5} torr. The workhorse loop generated 1.355 kW_e with a 1,033 K turbine inlet temperature. Disassembly and inspection showed all elements in excellent condition with no abnormal wear or degradation. At the time of termination of the DIPS program in 1978, the Closed Brayton Cycle was judged to be able to meet all DIPS space power requirements.

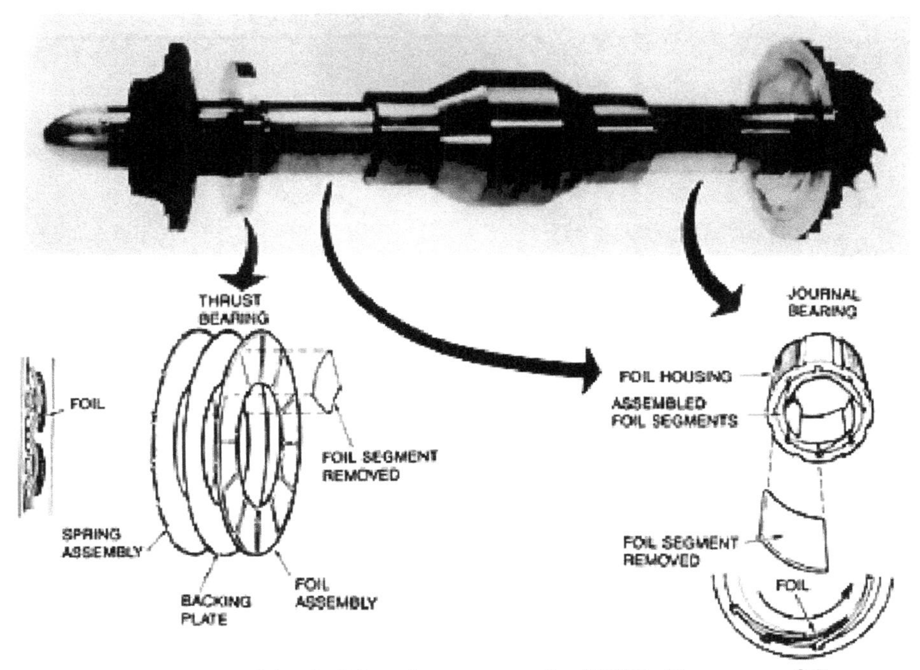

Fig. 9. Definition of the foil bearing system for DIPS. *Courtesy of Grumman.*

In the mid-1980's, space Brayton was revived as a component of the Space Station program. This was in the form of a 25-kW_e Solar Dynamic Power Module. In the 1994-1998 time period, as part of the Solar Dynamic Ground Demonstration (SD GTD), a 2-kW_e end-to-end system test was performed in a thermal-vacuum facility with a solar simulation. The system used the Mini-BRU components and compiled over 800 hr of operation.[16, 17]

How to handle redundancy to protect against single failure points of active components and reliability of the power system for long-duration missions has been a concern for dynamic isotope systems. For a DIPS-PCA (Power Conversion Assembly) subsystem to reach a reliability goal of 0.996 will require two or more PCA units.[18] The current-assessed single PCA operating reliability is 0.908. Assuming conservative 0.99 startup and switching reliabilities, various configurations of two and three PCAs were evaluated. The results, shown in Fig. 10, indicate that one unit operating and one in standby meets the reliability goal while weighing less then a three PCA configuration. The reliabilities for various configurations are summarized:

- Single PCA unit 0.908
- Two operating 0.991
- One operating and one standby 0.993
- Three operating 0.999
- One operating and two standby > 0.999

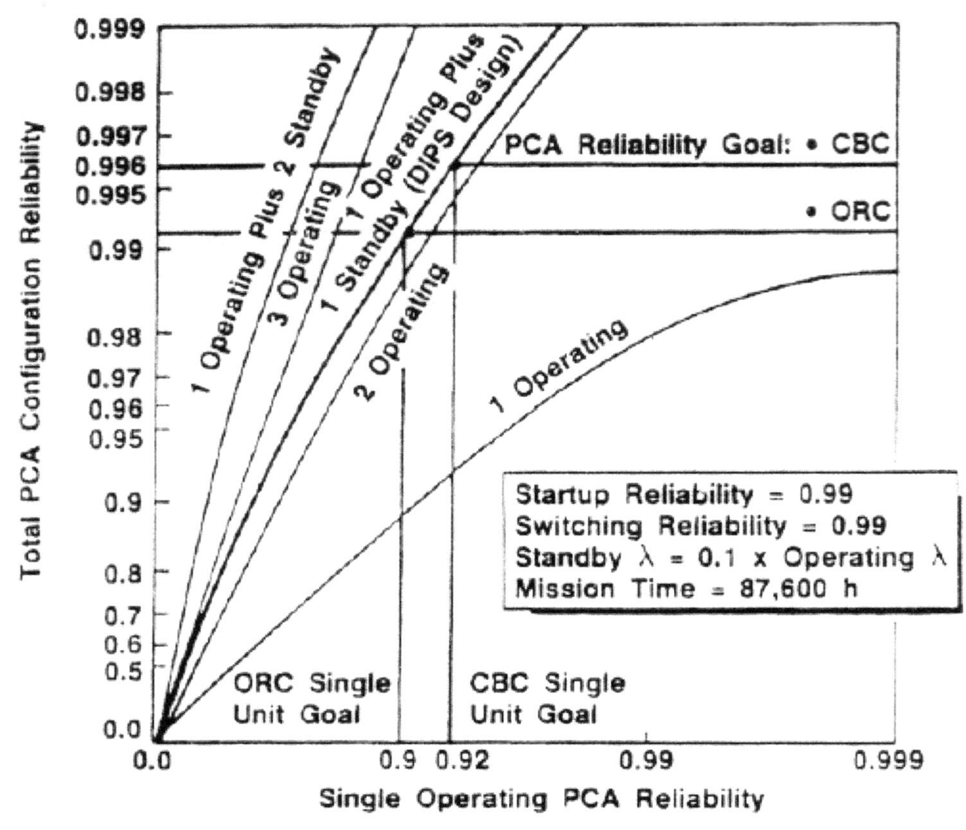

Fig. 10. Parametric impact of PCA operating and standby redundancy.

In the 1988 to the 1991 time period, a DIPS program was performed for possible use with a radioisotope power source. A DIPS Integrated System Test (IST) had a goal to demonstrate DIPS integrated operation, demonstrate long-term containment boundary integrity (for example, no loss of HE / Xe working fluid), and demonstrate flight readiness.

In 1994, a design that incorporated fully redundant power turboalternator compressor (TAC) conversion system for interplanetary missions resulted in the development of a conceptual design. This design used the following assumptions and guidelines

- Applicability
  - Scalability by changes only in Heat Source Unit (HSU) length, recuperator size, and number of radiator heat pipes
  - All components usable on Lunar/Mars surface
- Outer-planetary mission configuration
  - Fully redundant power conversion systems
  - 12-year full power life
- Superalloy gas containment boundary
  - No refractory materials outside of the HSU
  - Turbine inlet temperature = 1130 K
- Aluminum honeycomb heat pipe radiator
  - Two-sided radiation to space
  - Z-93 optical surface (T_{sink} = 167 K)
  - Direct coupled redundant gas manifolds
- Preserve simplicity of the cycle
  - Alternator cooling provided by cycle working gas
  - No auxiliary cooling loop

Table 1 summarize the key parameters for a 500 W_e and 1.0 KW_e system. Both power levels use the same TAC, but the 500 W_e unit uses fewer GPHS modules, scales down the recuperator size, and scales down the number of radiator heat pipes and area. The 500 W_e system has a specific power of 2.0 W_e / kg and the 1.0 kW_e unit a specific power of 3.1 W_e / kg. Fig. 11 shows a possible flight configuration. A flow diagram for DIPS IST is shown in Fig. 12.[19]

Table 1. DIPS Performance for 500 kW_e and 1.0 kW_e with redundant power conversion elements.

	500 W_e EOM*	1.0 kW_e EOM
System Mass, kg		
Heat Source Unit (HSU)	39	59
Turbo-alternator Compressor (TAC) (2)	37	37
Recuperator (2)	47	61
Manifold (2)	26	35
Radiator	15	22
Ducting (2)	19	19
Power Processing and Control (PP&C) (2)	63	86
Total, kg	246	319
W_e/kg	2.0	3.1
Radiator Area, m^2	4.7	6.5
Radiator Area, ft^2	50.8	69.8
Number of GPHS modules	13	21
Net Thermal Efficiency	0.170	0.211
Recuperator Efficiency	0.962	0.961

Table 1. DIPS configuration, assumptions, and guidelines.

At the time of termination of the DIPS program, the Closed Brayton Cycle was judged to be able to meet all DIPS space power requirements. Table 2 summarizes the technology base for Brayton Closed Cycle systems.[20]

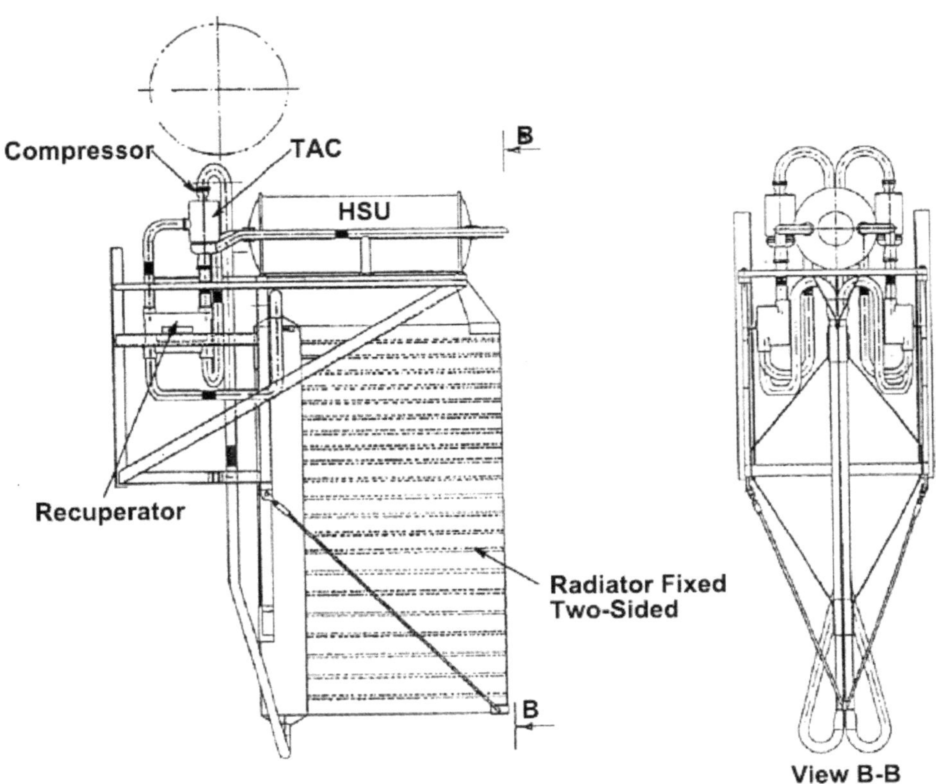

Fig. 11. DIPS flight configuration.

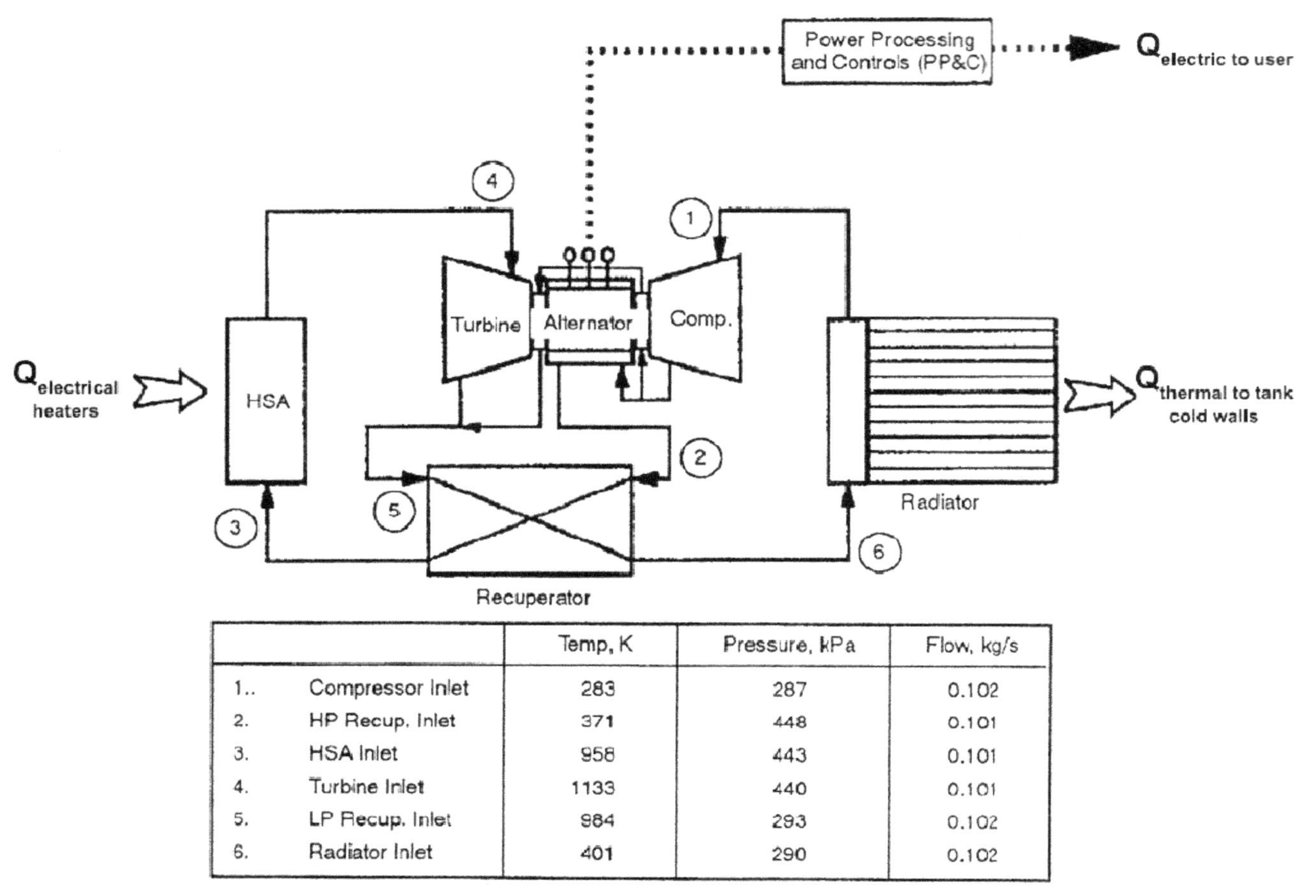

		Temp, K	Pressure, kPa	Flow, kg/s
1..	Compressor Inlet	283	287	0.102
2.	HP Recup. Inlet	371	448	0.101
3.	HSA Inlet	958	443	0.101
4.	Turbine Inlet	1133	440	0.101
5.	LP Recup. Inlet	984	293	0.102
6.	Radiator Inlet	401	290	0.102

Fig. 12. DIPS Integrated System Test state point diagram.

Table 2. Power conversion technology base for closed Brayton cycle components.

Component	Material	Temperature (K)	Pressure (kPa)	Size (diameter) (cm)	Speed (rpm) (000)	Hours Experience
Turbine	Inconel 713 LC Inconel 617	1030 to 1200	560 to 720	7.6 to 20	36 to 60	200 million in APUs 50,000+ on BiPS, BRU (Alt. Conditions)
Bearings	15-5PH	505	101-690	5.1	30 to 60	85 million, 20,000 each
Alternator Rotor	4340 and Inconel 718	415	183	8.8	36	51000
Compressor	Ti-6Al-4V/347SS	505	730	17	60	200 million on APUs, 50,000+ on BIPS, BRU (Alt. Conditions)
Recuperator	AISI 409SS	865	965 to 1200			768000

Rankine Cycle Converters For Radioisotope Generators

Rankine Cycle Principles

The Rankine thermodynamic cycle is named after the Scottish engineer William Rankine (1820-1872). He made fundamental contributions to engineering thermodynamics, especially with respect to the steam engine. The basic Rankine cycle is a form of heat engine that includes a working fluid having two-phases as part of the cycle.

In simple terms, the Rankine cycle heat engine is composed of a heat source, in this case the radioisotope fuel; a turbine-generator where thermal energy is converted to electric power; a heat rejection radiator where the fluid undergoes a phase change and waste heat is rejected to space; and a pump to increase working fluid pressure. In thermodynamic terms, the Rankine cycle (Fig. 13) working fluid undergoes an isentropic expansion in the turbine from state 1 to state 2. The "wet" liquid-vapor mixture then experiences a phase change to a saturated liquid as it undergoes isothermal heat rejection in the condensing radiator (state 2 to state 3). This is followed by isentropic compression in the pump from state 3 to state 4. Here, the liquid is assumed to be incompressible and actually goes from the saturated liquid state (3) to the sub-cooled liquid state (4). Finally, the working fluid experiences constant pressure heat addition from state 4 to state I. At state 4' sensible heat addition gives way to phase change--so that, as more thermal energy is added, the working fluid gradually changes from a saturated liquid (point 4') to a saturated vapor (point 1) of 100 percent quality. This phase change process occurs at both constant temperature and constant pressure.

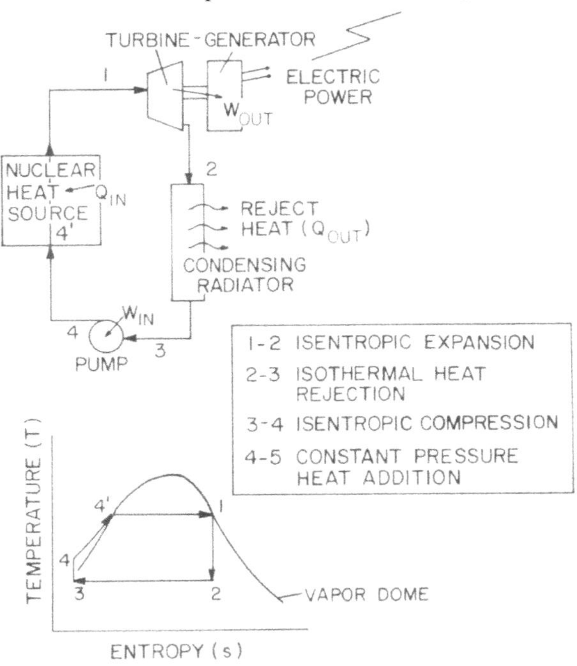

Fig.13. Basic Rankine cycle (ideal).

The quality (x) of a liquid-vapor mixture is defined in thermodynamics as the fraction of mass of a working fluid in the vapor phase. Similarly, the wetness of a mixture ($1 - x$) describes a liquid-vapor mixture with a quality less than 100 percent. The thermodynamic properties of a two-phase mixture are designated in terms of the following specific enthalpies:

$$h_{\text{mix}} = (1 - x)h_f + xh_g \tag{9}$$

where h_{mix} is the enthalpy of the mixture (J / kg)

h_f is the enthalphy of the saturated liquid (subscript "f" for liquid phase) (J / kg)

h_g is the enthalphy of the saturated vapor (subscript "g" for vapor phase) (J / kg)

x is the quality of the mixture.

A subcooled or compressed liquid is one that exists at a temperature lower than the saturation temperature, corresponding to its pressure; while a superheated vapor is a vapor existing at a temperature greater than the saturation temperature, corresponding to its pressure.

Neglecting changes in the potential and kinetic energy of the working fluid as it goes around the basic Rankine cycle, the thermal efficiency is

$$\eta_{th} = \frac{(\text{work out})_{\text{turbine}} - (\text{work in})_{\text{pump}}}{\text{heat added}} . \tag{10}$$

Assuming ideal components throughout the system yields

$$\eta_{th} = \frac{(h_1 - h_2) - (h_4 - h_3)}{(h_1 - h_4)} . \tag{11}$$

The thermal efficiency of the Rankine cycle can be improved by superheating the working fluid, as shown in Fig. 14. Here, the constant pressure heating of the working fluid is continued past the saturated vapor state (state I) and into the superheated vapor region (state 1').

The thermal efficiency of the basic Rankine cycle can also be improved using the reheating process shown in Fig. 15. The working fluid is expanded in the first turbine stage (from point l' to 2) until it reaches pressure p_2 (on the saturated vapor line). Then, the working fluid is returned to the nuclear power source where it is reheated at constant pressure to temperature T_2 in the superheated vapor region (state 2). Next, it undergoes expansion to state 3 in the second stage of the turbine.

Superheat and reheat reduce the amount of moisture in the working fluid entering the turbine. This prevents liquid droplet erosion of turbine blades and raises the average temperature at which thermal energy is added to the working fluid. Both techniques are used to improve overall Rankine cycle efficiency.

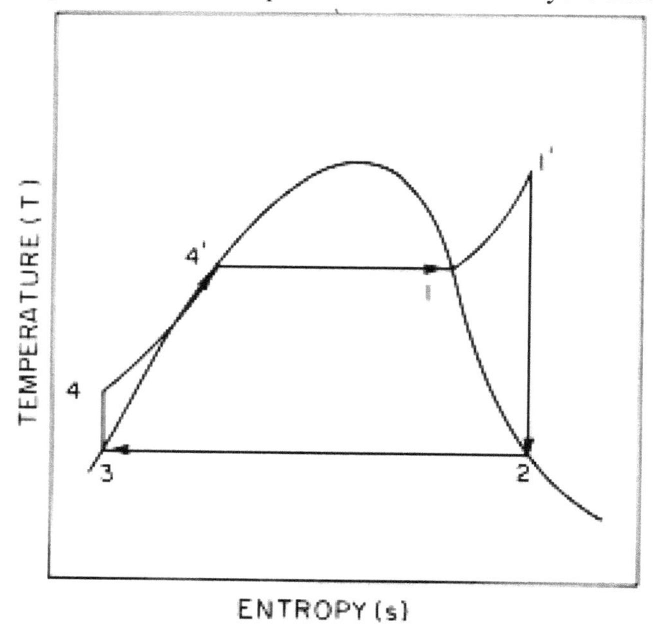

Fig.14. Rankine cycle with superheat.

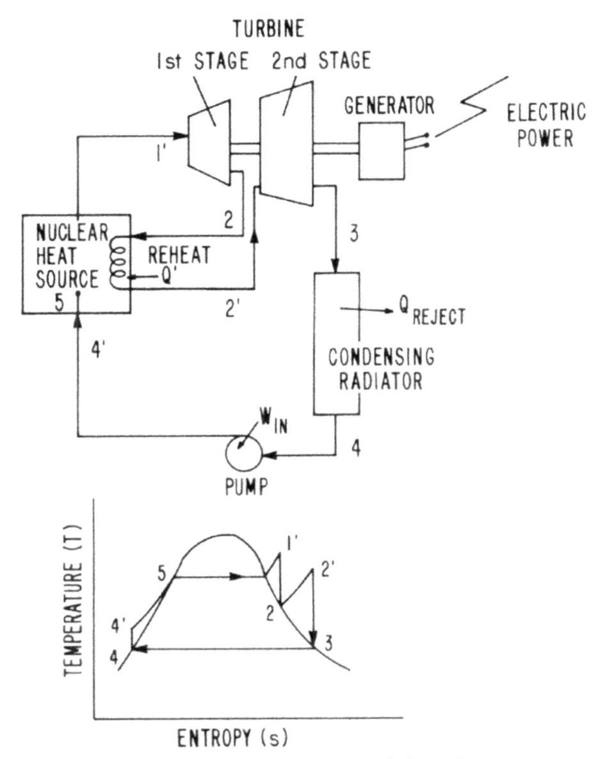

Fig. 15. Rankine cycle with reheat.

Regeneration also can be used to improve the overall efficiency of the Rankine cycle. As shown in Fig 16, when the working fluid expands in the turbine some of its thermal energy content is used to preheat the liquid-phase working fluid before it enters the radioisotope nuclear heat source. This technique allows a close approach to the ideal Carnot cycle efficiency, since thermal energy transfer in the overall cycle now occurs isothermally. An ideal turbine as a power producing device is not an ideal heat exchanger. Therefore, many engineering trade-offs are made in practical Rankine cycle applications, including the use of superheat, reheat, and regeneration.

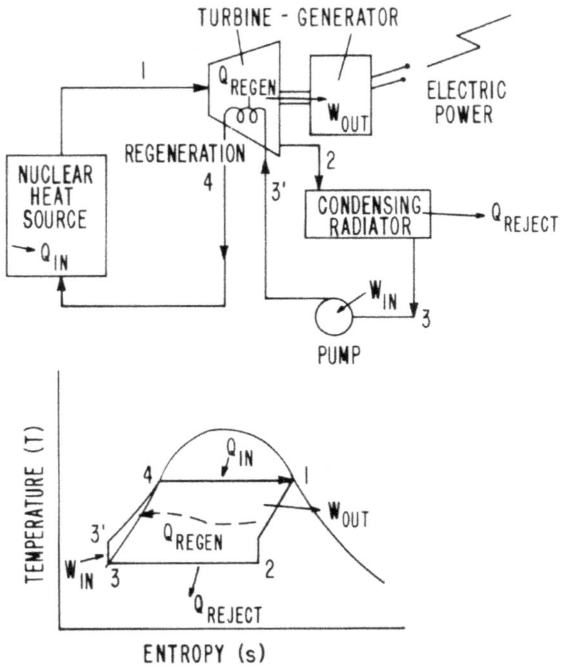

Fig. 16. Ideal Rankine cycle with regeneration.

Dynamic Isotope Power System (DIPS) Development

The Dynamic Isotope Power System (DIPS) program goals were to define flight conceptual designs in the 0.5 to 2 kW_e range with a reliability of 0.95, lifetime of 7 years, and to ground demonstrate a 1.3 kW_e unit. One concept, Kilowatt Isotope Power System (KIPS), used an organic working fluid Rankine cycle system. The KIPS concept was selected in the DIPS program for further development. It reached the stage where a ground demonstration unit was completed, but the program itself has been inactive since December 1980.

Fig. 17. is a schematic diagram of the DIPS power plant, Fig. 18 shows the overall power system, and Fig. 19 illustrates typical operating conditions. The organic working fluid is an organic eutectic mixture of biphenyl and biphenyl ether called Dowtherm A. The heart of the DIPS system is the turbine-alternator-pump power conversion assembly. These are integrated onto a common shaft (see Fig. 20). This single unit is supported by tilt-pad thrust and journal bearings, lubricated by the Dowtherm A working fluid. No physical contact occurs during system operation between the rotating shaft and the bearings, except for a few seconds at startup.

The DIPS has two major flow paths (see Fig. 17). At the pump discharge, the Dowtherm A organic fluid is a liquid. Approximately 10 percent of the working fluid moves through the inside of the tubes in the regenerator, extracting heat from and cooling the turbine discharge vapor. The working fluid then passes through the spiral tubes surrounding the isotope heat source. Here, the Dowtherm A is vaporized and heated to its peak temperature of 630 K. Next, the vapor is expanded through the turbine. The turbine removes energy and converts it to shaft power to drive the alternator and pump. The vapor leaves the turbine and passes over the external surfaces of the regenerator tubes. This preheats the pump discharge flow and thus enhances the system efficiency. The vaporized working fluid is then condensed on the sub-cooled radiator flow in the jet condenser. It then proceeds to the accumulator, returning to the pump to complete the Rankine power conversion cycle.

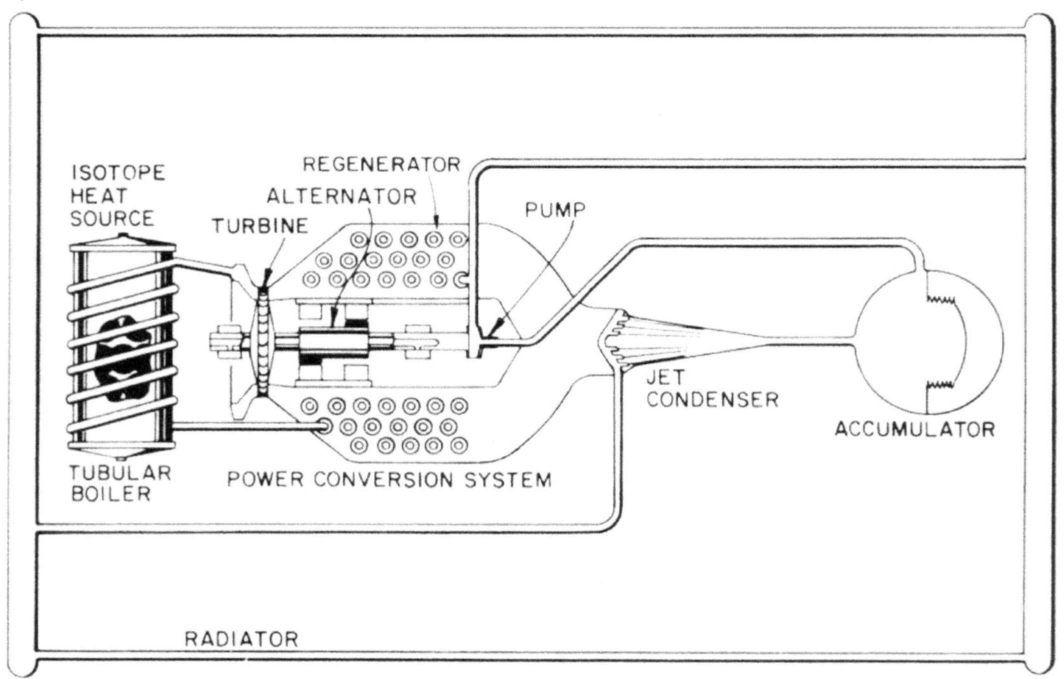

Fig. 17. Schematic of Dynamic Isotope Power System (DIPS) power plant.[21]

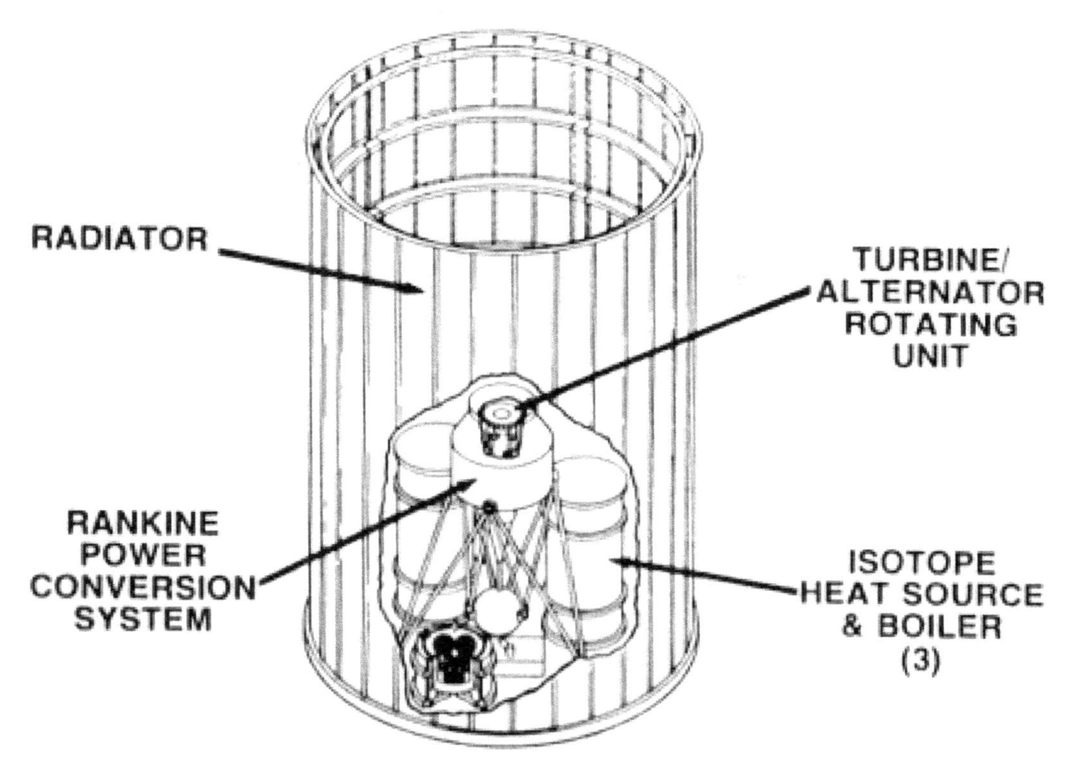

Fig. 18. Kilowatt Isotope Power System (KIPS).

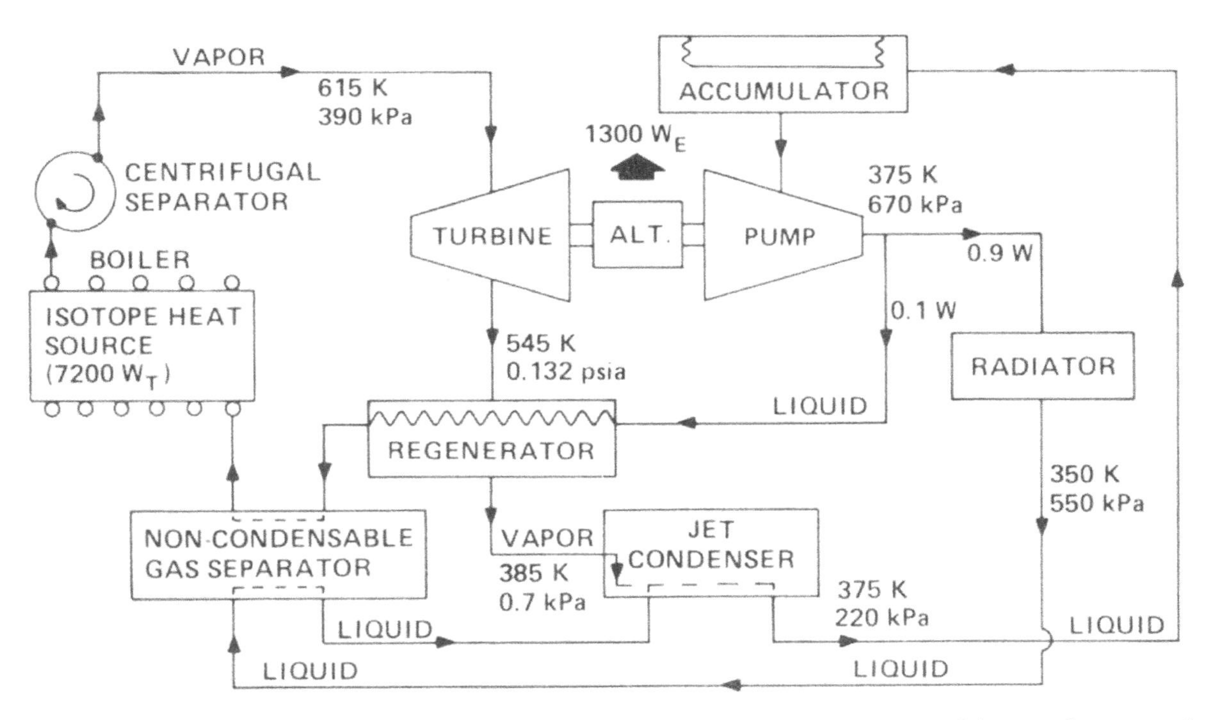

Fig. 19. Representative operating conditions for radioisotope organic Rankine cycle generator system; the working fluid is Dowtherm A.[22]

In the second major working fluid path described in Fig. 19, the other approximately 90 percent of the organic working fluid flow passes through the radiator as a single phase fluid (liquid). There is no condensation or two-phase flow in this radiator. The working fluid is cooled in the radiator as a liquid, rejecting the waste heat extracted from the vaporized working fluid in the jet condenser. The liquid Dowtherm A then exits from the radiator to a shower-head type fixture in the jet condenser. The multiple liquid streams generated in the shower head fixture provide the surface area and heat capacity needed to condense the vapor as it exits from the power conversion system regenerator. The jet condenser design for the DIPS is fully effective in the microgravity environment of outer space. Also, it will operate at acceleration levels up to 30 g's, as may be encountered during launch and orbital insertion maneuvers. Ancillary temperature regulation valves and electrical controls (not shown in the figures) maintain the proper relationships of the working fluid temperatures and regulate the electric power output of the system. Some of the characteristics of the DIPS flight development system are summarized in Table 3.

Table 3. Dynamic Isotope Power System (DIPS) flight development system characteristics.

Rated Output Power, EOM (W_e)	1,300
Optional Ratings (W_e)	500 to 2,000
Input Power @ 1,300 W_e	7,200
Overall Thermal Efficiency[1]	18.1% rectified to 28 VDC
Peak Working Fluid Temperature (K)	630
Total Weight[2] (kg)	215
Envelope Dimensions for 1,300 W_e BOM[3] (cm)	
Diameter	132
Length	24
Design Point Output Voltage (VDC)	24 ±2
Response--Variable Output Power--0 to 100%	Milliseconds
Partial Output Power Capability	On Pad Launch to Parking Orbit. Orbital Transfer to Final Orbit
Output Power Capability--0 to 100% Spacecraft	Spin Stabilized
Lifetime (years)	7
Capable of Stable Operation with Unbalanced Solar Input	
Resistant to Natural and Induced Radiation	

[1] With RTG topping system efficiency can be increased to 22.1% or higher.
[2] Based on updated flight system design utilizing conventional materials, weight reductions such as alternate radiator configuration and converter rotating unit housing machining, possible to approach the 204 Kg goal.
[3] Flexible in design for both size and shape to fit the spacecraft.

The ground demonstration unit was run for 4,298 h. After inspection, all the hardware was found to be in excellent condition except for some slight burnishing of the radial bearing. An endurance test of 2,002 h was successfully run. In all, 11,146 h of testing was accumulated in the ground demonstration program. It was concluded that KIPS was inherently capable of 60,000 h of operation.

KIPS used Dowtherm A as the working fluid and a Combined Rotating Unit speed of about 577 Hz (34,600 rpm). The overall system efficiency achieved in the ground demonstration phase was 16.6% DC and 18.5% AC.[23] A lower system mass with efficiencies on the order of 24% can be achieved with. a higher Combined Rotating Unit speed of around 1,000 Hz, Toluene as the working fluid and a turbine inlet temperature of about 670 K[24]

Table 4 summarizes the power conversion data technology base.[25]

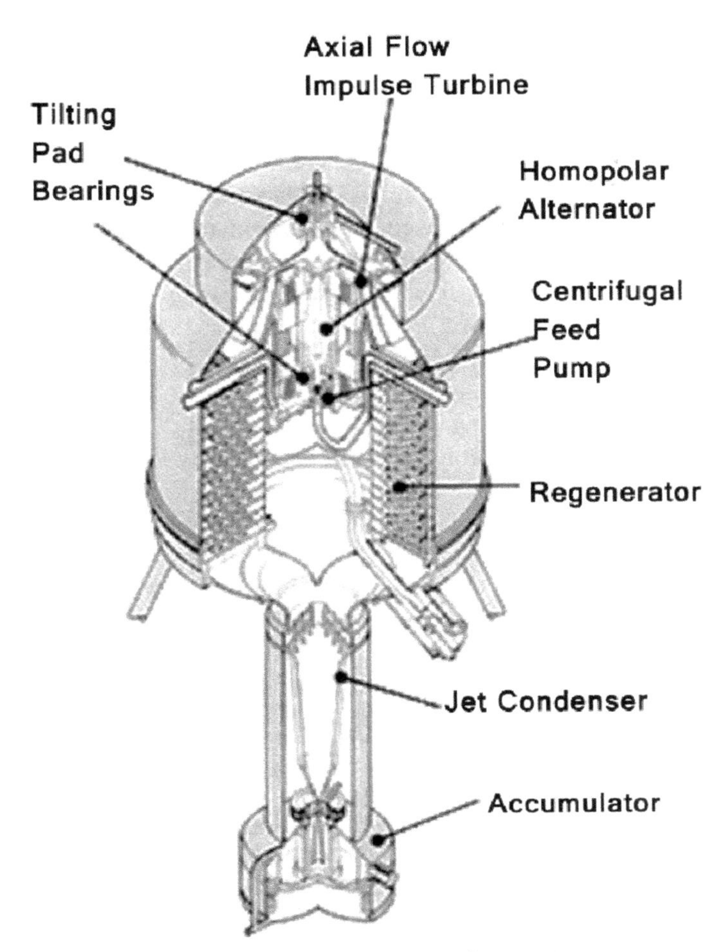

Fig. 20. DIPS power conversion system.

Table 4. Power conversion technology base for Organic Rankine Cycle components.

6 kWe DIPS						Applicable Equipment Experience					
Component	Material	Temper -ature (K)	Pressure (kKa)	Size (diameter) (cm)	Speed (rpm) (000)	Material	Temper-ature (K)	Press ure (kPa)	Size (diameter) (cm)	Speed (rpm) (000)	Hours Experience
Turbine	Inconel 718	670	4870	9.4	60	Same	715	4420	25.4	55	>100,000
Bearings	M-50/Ag Plated	340	24	1.3	60	Same	350	7-3440	2.8	50	>100,000
Alternator	AMS6302	360	5660	5.1	60	Same	360	345-3790	10.7	50	>100,000
Feed Pump	17-4PH	350	5750	3.6	60	Same	355	4620	3.3	50	>100,000
Rotating Fluid Management Device	6061 Al/17-4PH	340	24	19.1	4.0	Same	270-320	345-1930	19.9	1378	12,000
Regenerator	347SS	505	5280	29 x 20 L		Same	520	660	60x46 L		>100,000

Several advances have been made since the KIPS program ended. The converter can be operated in a supercritical fluid cycle which allows heat addition to the organic fluid without a boiling zone in the vaporizer. This means that the only place where two-phase fluid management is needed is on the condensing side and this is accomplished in zero-G with a rotary fluid management device and a shear-flow-controlled condenser design. The rotary fluid management device and shear-flow-condenser were successfully tested using freon in 54 KC-135 flight simulations of microgravity.

Stirling Cycles

***Principles of Stirling Cycles*[26]**

To achieve a totally reversible heat engine involves devising a means by which all thermal energy transfer to and from the system takes place both isothermally and reversibly. Robert Stirling (1790-1878) designed such a cycle in the early 19th century. In fact, the Stirling engine is the earliest example of a reversible heat engine. Robert Stirling, along with his brother James, used regeneration in developing this engine. Stirling engines enjoyed considerable success as quiet pumping engines. However, with the development of more compact internal combustion engines, Stirling engines fell into disuse. Space power applications have revised interest in the Stirling cycle.[27]

A heat engine may incorporate a component called an ideal regenerator. Through the regenerator, heat is alternately stored and then recovered, reversibly. To accomplish this, the working fluid transfers heat on a temporary basis, while dropping in temperature from some upper value (T_{high}) to some lower value (T_{low}). Inherent in this ideal regenerator concept is the assumption that heat transfer occurs reversibly--that is, there is no temperature difference between the heat absorbing regenerator material (such as wire mesh or tiny thin-walled tubes) and the working fluid at any point where they are in thermal contact (see Fig. 21). Then, when the working fluid passes back through the regenerator, entering at T_{low}, it recovers the heat which was originally stored and leaves at T_{high}. Since the regenerator is in exactly the same state after completion of a cycle as it was before, it is considered as a component of the heat engine and not as a part of some external heat source or sink. In the Stirling engine this reversible heat exchange in the regenerator occurs at constant volume.

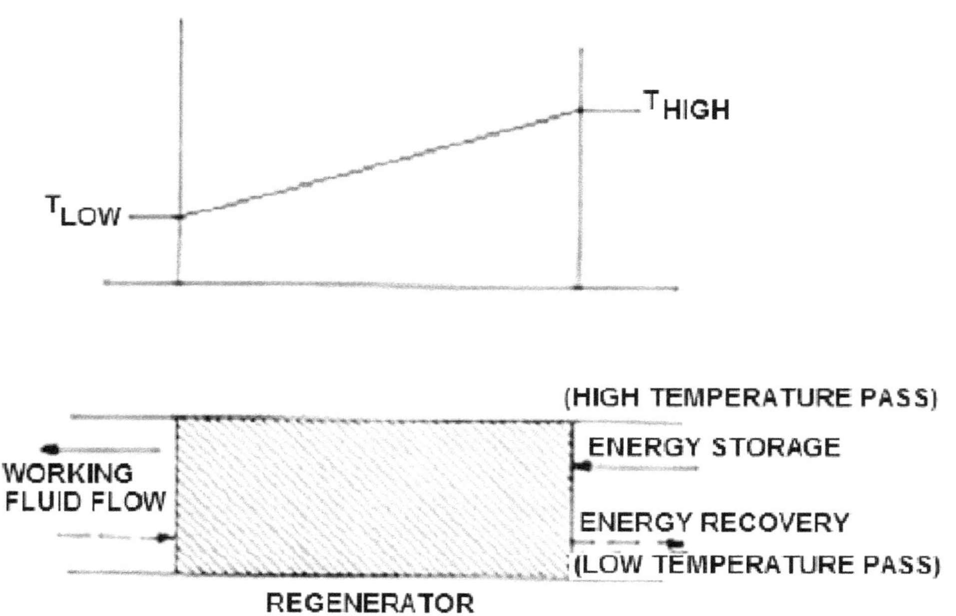

Fig. 21. Flow of a working fluid through an ideal regenerator.

Fig. 22 illustrates the basic Stirling cycle with an ideal gas working fluid. From state 1 the gas, initially at the lower temperature limit, recovers the stored thermal energy from the regenerator in a reversible, constant volume process. This continues until the working fluid reaches the upper temperature limit, state 2. Then, from state 2 to state 3 the working fluid experiences reversible heat addition from the external heat source (also at T_{high}) and expands isothermally to state 3. From state 3 to state 4, the working fluid again interacts with the regenerator in a constant volume process. This time, however, thermal energy is transferred from the working fluid to the regenerator. The ideal gas goes from T_{high} at state 3 to T_{low} at state 4. An isothermal compression takes place from state 4 to state l. During this final portion of the Stirling cycle, the working fluid rejects heat to an external sink that is also at T_{low}. The thermal efficiency of the ideal Stirling cycle is

$$\eta_{Stirling} = 1 - T_{low}/T_{high} \tag{12}$$

where T_{low}, T_{high} are absolute values of temperature (K). Equation 12 is the identical mathematical expression for the thermal efficiency of the Carnot cycle operating between the same temperature limits.

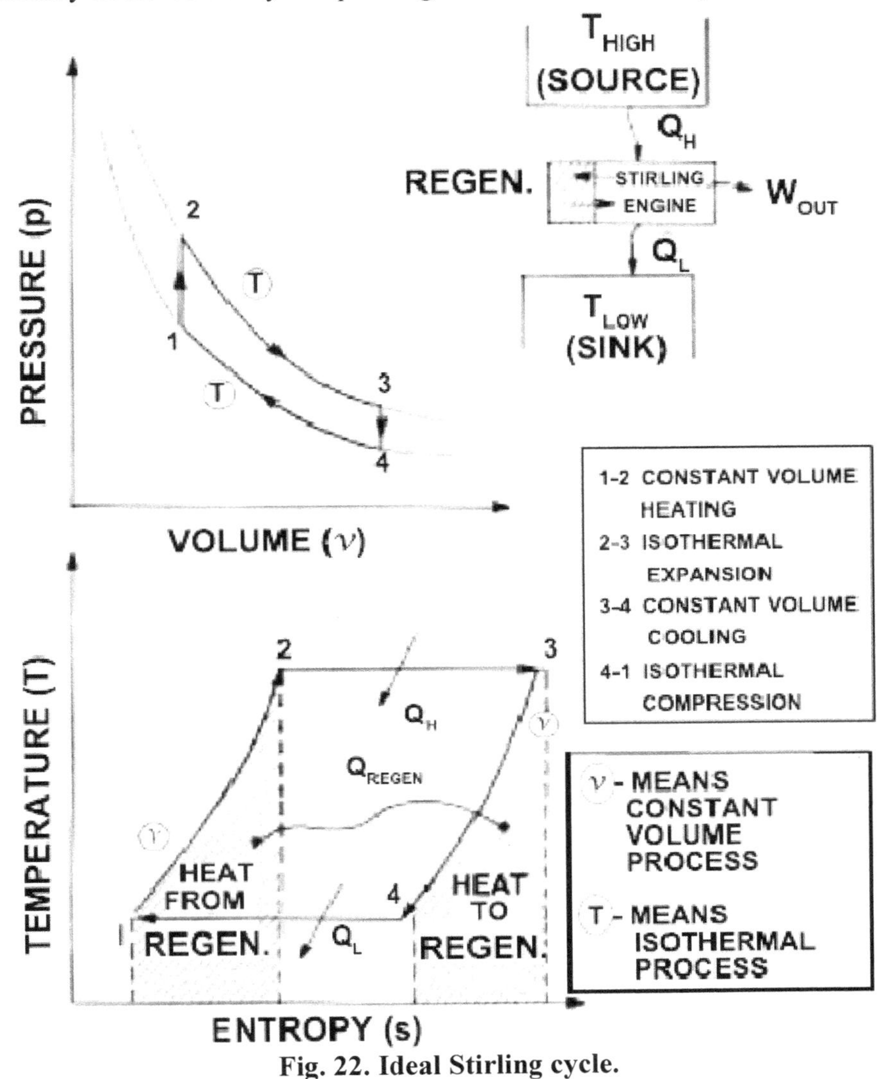

Fig. 22. Ideal Stirling cycle.

The free piston Stirling engine (FPSE) of interest in space applications is a thermally driven mechanical oscillator, using the Stirling cycle and deriving its output power from heat flow between a source and a sink.[28, 29] The FPSE displacer motion is produced by gas pressures rather than by mechanical linkages. This type of engine operates at the highest device efficiency of all known heat engines and is uniquely suited to drive direct-coupled reciprocating loads (such as linear generators) in a hermetically sealed configuration and without requiring high pressure shaft seals or contaminating lubricants.

The Free-PistonStirling Engine (FPSE) consists of three basic components (see Fig. 23): a power piston (labelled 'piston' on the figure), the displacer piston and a sealed cylinder. The displacer and the power piston assembly are the only two moving parts in the FPSE engine-alternator unit. The displacer rod passes through the power piston and is in communication with the bounce space. The bounce space acts as a constant pressure chamber. The power piston assembly is composed of the power piston directly connected to the alternator armature. This eliminates the need for converting reciprocating motion to rotary motion and importantly the need for lubricating such a Stirling device. The assembly is supported by two pressurized (hydrostatic) helium gas bearings. Long-term reliability is inherent in this design, because compressor inlet and discharge flow is controlled by means of fixed ports rather than valves.

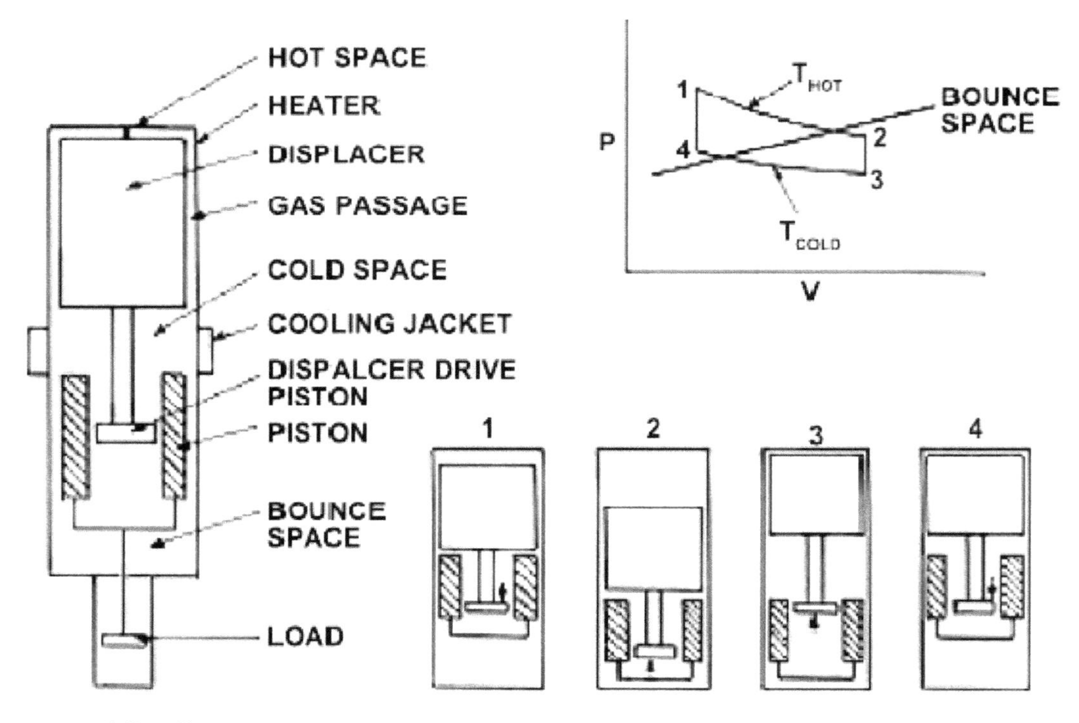

1-2 PISTON EXPANDS WORKING GAS DISPLACER ON PISTON
2-3 PRESSURE IN BOUNCE SPACE GREATER THAN PRESSURE IN WORKING SPACE, FORCING DISPLACER TOWARD HOT SPACE, WORKING GAS MOVED INTO COOLED SPACE, PRESSURE DROPS
3-4 PISTON DRIVER INTO WORKING SPACE BY HIGHER BOUNCE SPACE PRESSURE
4-1 DISPLACER DRIVEN TOWARD COLD SPACE BY WORKING SPACE PRESSURE HIGHER THAN BOUNCE SPACE PRESSURE

Fig. 23. Operational principles of free piston Stirling engine (FPSE).

Advanced Stirling Radioisotope Generator (ASRG)

The ASRG (Fig. 24) will use two Stirling engines with one GPHS heat source per engine. These are coaxially aligned and in opposed operation to minimize piston-induced vibration. The generator is designed to produce 151 W_e power at the beginning-of-mission and to have a design life of 14 years plus three years in storage. The ASRG include:

- an integrated, single fault-tolerant controller based on active power factor correction,
- autonomous operation and fault isolation from the space vehicle,
- one-half generator operation capability in case of an engine failure, and
- operation in vacuum of deep space and in planetary atmospheres.

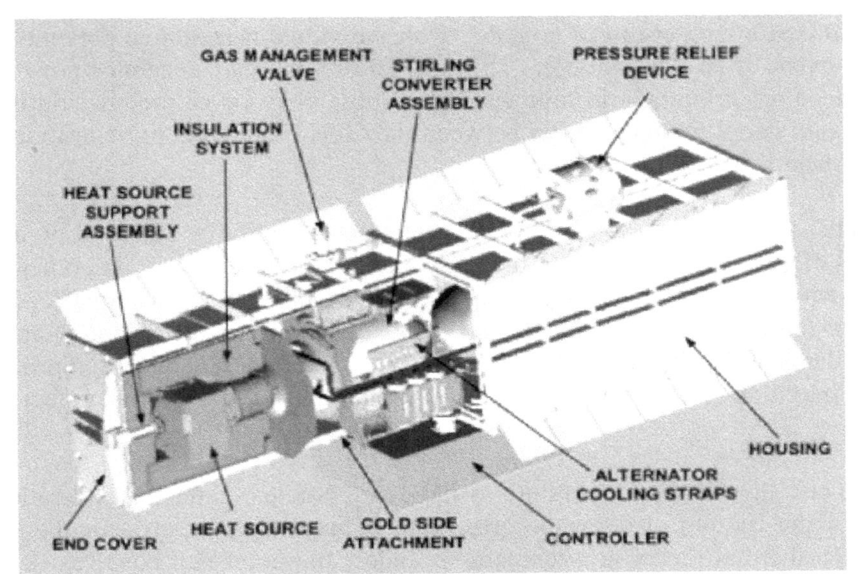

Fig. 24. Advanced Stirling Radioisotope Generator.

A schematic of the Stirling convertor is shown in Fig. 25.[30] with an external view in Fig. 26. The Advanced Stirling Converter currently part of the Engineering Unit uses a heater head made of Inconel 718. This limits the hot-end temperature to about 925 K. With the use of MarM-247 superalloy (% weight Mo 0.65, W 10.0, Ta 3.3, Ti 1.05, Cr 8.4, Co 10.0, Hf 1.4, Zr 0.055, B 0.15, balance Ni) for the heater head material, the allowed operating temperature can be increased to 1,125 K. With this hot-end temperature and a temperature ratio of 3:1, the efficiency of the convertor can reach nearly 40% from heat input to alternating current output from the alternator.[31]

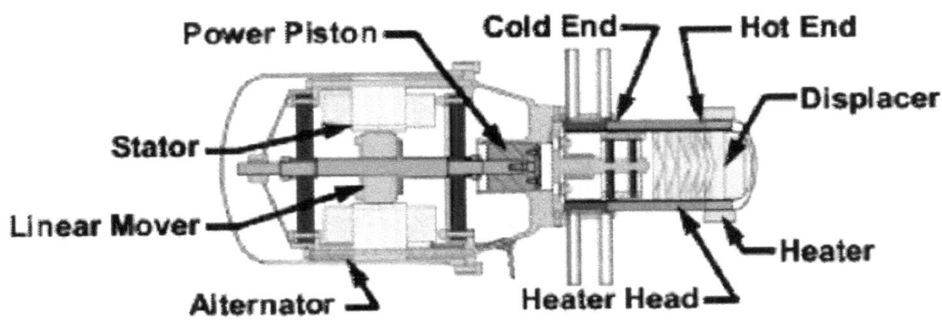

Fig. 25. Stirling Converter Assembly.

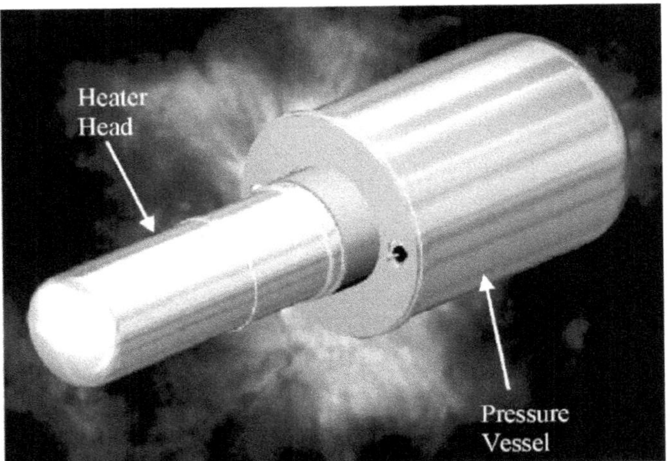

Fig. 26. External view of the Advanced Stirling Converter (ASC-2)[32]

The approach of dual ASCs with each one coupled to an individual heat source prevents fault propagation in the generator in the event of an ASC failure. This still allows one-half generator power output. Using end mounting eases the loading of the radioisotope general purpose heat source modules during generator fueling. It is necessary to ensure good thermal contact between the GPHS and ASC hot end to minimize temperature gradient and thermal heat loss.[33]

The primary structure is a beryllium housing. The square cross-section is stiffened by axial and lateral ribs. The housing consists of two halves in order to facilitate integration of the ASCs and other internal components during generator assembly. Beryllium fins provide additional radiative surface area for waste heat rejection. The fins are bolted to four corners along the length of the housing rather than being attached by brazing. This mitigates accidental damage to the fins during fabrication and generator assembly and processing. The fins are made from 0.060-in beryllium--making them very light weight and have a high view factor from both sides to space.

A failed ASC will cause the GPHS temperature to increase. Nuclear safety considerations requires that the structural integrity of the iridium cladding be preserved. Thus, a key criteria for thermal insulation is the ability to reject GPHS thermal power at a temperature below the point that could cause grain growth in the iridium clad encapsulating the radioisotope fuel pellets. Proposed insulator materials--Microtherm HT, Min-K and Microtherm HT/Aspen aerogel combinations--achieved heat rejection capability below the grain-growth temperature limit by their physical shrinkage and increased thermal conductivity at above operating temperatures.

During ground operations, an externally mounted gas management valve (GMV) is used to maintain the interior of the generator inert gas (argon) above atmospheric pressure. This prevents oxidation of graphite and refractory metal components in the Electrical Heat Source (EHS) at elevated temperatures. During launch, the barometrically operated pressure relief device (PRD) will be actuated to vent the inert gas to space. For Mars surface application, a vent tube connecting the generator housing to the activated PRD is sized to provide a reduced gas conductance from the Mars atmosphere to the interior of the generator. This reduces the mass loss of the GPHS graphite aeroshell from CO_2 gasification to an insignificant level for the 14-year mission life.

The Advanced Stirling converters have continued to evolve. Table 5 list different versions of ASC machines. The FTB (Frequency Test Bed) was used to investigate performance at power levels about 80 W_e with convertors operating at high frequency (> 100 HZ). These achieved a record setting 36% efficiency at a temperature ratio of 3.0. The ASC-1 units included the use of higher temperature heater head materials capable of > 17 y lifetimes. The ASC-2 further evolved the convertors with lighter weight units being the objective.[34]

Table 5. Different versions of ASC machines.[35]

	Units	Head Material	Head Temperature (C)	Hermetic	Additional Design Evolution
FTB	2	Stainless Steel	650	No	
ASC-1a	2	MarM-247	850	No	
ASC-1b	2	MarM-247/IN-718	850	No	Inertia welded head
ASC-0	2	IN-718	650	Yes	Brazed displacer dome/body joint
ASC-1HS	2	MarM-247/IN-718	850	Yes	Improved piston/gas bearing/center port configuration
ASC-E	3	IN-718	650	Yes	Improved piston sensor, Inconel rejector pressure wall for reliability
ASC-2	4	MarM-247/IN-718	850	Yes	Diffusion bonded external acceptor

The ASC convertors are currently being run to accumulate hours and operating experience as shown in Table 6. The Engineering Unit will incorporate a convertor with an Inconel displacer and operate at 925 K. Therefore, priority has been given to operating at these conditions.

Simulated launch load vibration testing in both axial and lateral directions indicate that the generators meet Qualification level and Qualification plus 3 dB levels. The result are given in Table 7.

Table 6. Accumulated runtime on ASC-1 convertors.

ASC Accumulated Hours** as of April 16, 2007						Thermal cycles to
	Total	≥650 C	≥750 C	≥800 C	850 C	≥650 C
ASC-1 #1	126.0	97.5	46/5	46.5	44.5	35
ASC-1 #2*	86.0	64.5				17
ASC-1 #3	279.0	177.1	90.0	82.5	76.0	57
ASC-1 #4	135.0	83.9	37.2	36.2	32.1	42
ASC-0 #1	1151.0	876.0				55
ASC-0 #2	1155.0	880.0				52
Total	**2932.0**	**2179.0**	**173.7**	**165.2**	**152.6**	**258**
* Engine 2 has Inconel displacer **Operating hours do not include time during heat leak tests						Thermal cycles include heat leak tests

Table 7. Launch vibration levels during testing at PWR (in both axial and lateral directions).

	Level	Time (minutes)
Workmanship	6.8 g_{rms} random	1
Flight	8.7 g_{rms} random	1
Qualification	12.3 g_{rms} random	3
Qualification + 3dB	17.5 g_{rms} random	1

Also under development is a higher-temperature ASC with a MarM-247 heater head and an 850 C (1123 K) temperature. This increases the ASRG specific power to ~ 8.4 W_e / kg and provides increased margin with the MarM-247 material at 850 C compared to the Inconel 718 material at 650 C. Table 8 shows a comparison of the two configurations.[36]

An electrically heated engineering model of the Advanced Stirling Radioisotope Generator(ASRG) has accumulated over 3,700 hours of duration testing as of June 2009. The hot-end operated at a temperature of 913 K (640 C). The test unit delivered a stable power of 135 W_e. The ASRG project has proceeded through a systems requirements review in May 2009 and a final design review is anticipated in 2010. The goal is to have a flight system available for a mission in late 2013.[37]

Table 8 Comparison of 650 and 850 C ASRGs.

Parameter	ASRG--650 C	ASRG--850 C
Power per ASRG (beginning-of-life (BOL)) W_e	143	~ 160
Power degradation, %/yr	0.8 (power decays with fuel decay)	
Mass per ASRG, kg	20.2*	~ 19
Dimensions, mm	Length:725 Width: 293 Height: 410	TBD
Number of GPHS Modules	2	2
Thermal power input (BOL), W_t	500	500
ASRG specific power, W_e/kg	7.0	~ 8.4
Conversion efficiency, %	28	~ 32
Controller	Single fault tolerant	
Operating environment	Vacuum and Mars atmosphere	
Life requirement	14 yr mission + 3 yr storage	

* Add 1.2 kg for spacecraft adapter for missions using heavy launch vehicles.

Advanced Thermoelectric Converter (ATEC)

Current systems using thermoelectric converters have an efficiency of 6.6% and 5.2 W_e / kg for GPHS-RTG and 6% efficiency and 2.8 W_e / kg for MMRTG. A program is underway to develop a new generation of thermoelectric converters that would improve performance in the important W_e / kg metric by about 57%. The primary near-term goals in this program, called Advanced Thermoelectric Converter (ATEC), are:[38]

- Demonstrate a 35 - 65% improvement (8 - 10% overall) in conversion efficiency over that of GPHS-RTG and MMRTG in an Electrical Performance Demonstrator. This could take the form of an 18 thermocouple unit.
- Produces a viable RTG design showing capability for 6 - 8 W_e / kg specific power, with a goal of achieving 8 W_e / kg.
- Demonstrate one year life for four-couple modules, and greater than one year life for unicouples.
- Using a combination of life prediction models and experimental data, show a 14 year life and 3 years storage capability.

The performance of thermoelectric energy conversion devices depends on the thermoelectric figure-of-merit (ZT) of a material defined as $ZT = S^2\sigma T / k$, where S, σ, k, are the Seebeck coefficient, thermal conductivity, and absolute temperature, respectively. The figure-of-merit, ZT, directly impacts device efficiency. Several thermoelectric materials with high figures-of-merit (or high ZT materials) that can operate at temperatures of 1,075 -1,275 K have been identified (see Fig. 27 [39]). This temperature range has been determined to provide the optimum converter performance. Candidate materials selected are: p-type Zintl ($Yb_{14}MnSb_{11}$), n-type Lanthanum Telluride ($La_{3-x}Te_4$), a nano-structure p-type SiGe and a mechanically alloyed n-type SiGe. For instance, for the lanthanum telluride system, a ZT value at 1275 K for the pure $La_{3-x}Te_4$ system for x values near 0.2 is approximately one.[40] This compares favorably with the SiGe thermoelectric material used in the GPHS-RTGs. Adding Yb appears to further increase the ZT value from 1 to 1.2 at 1,275 K.

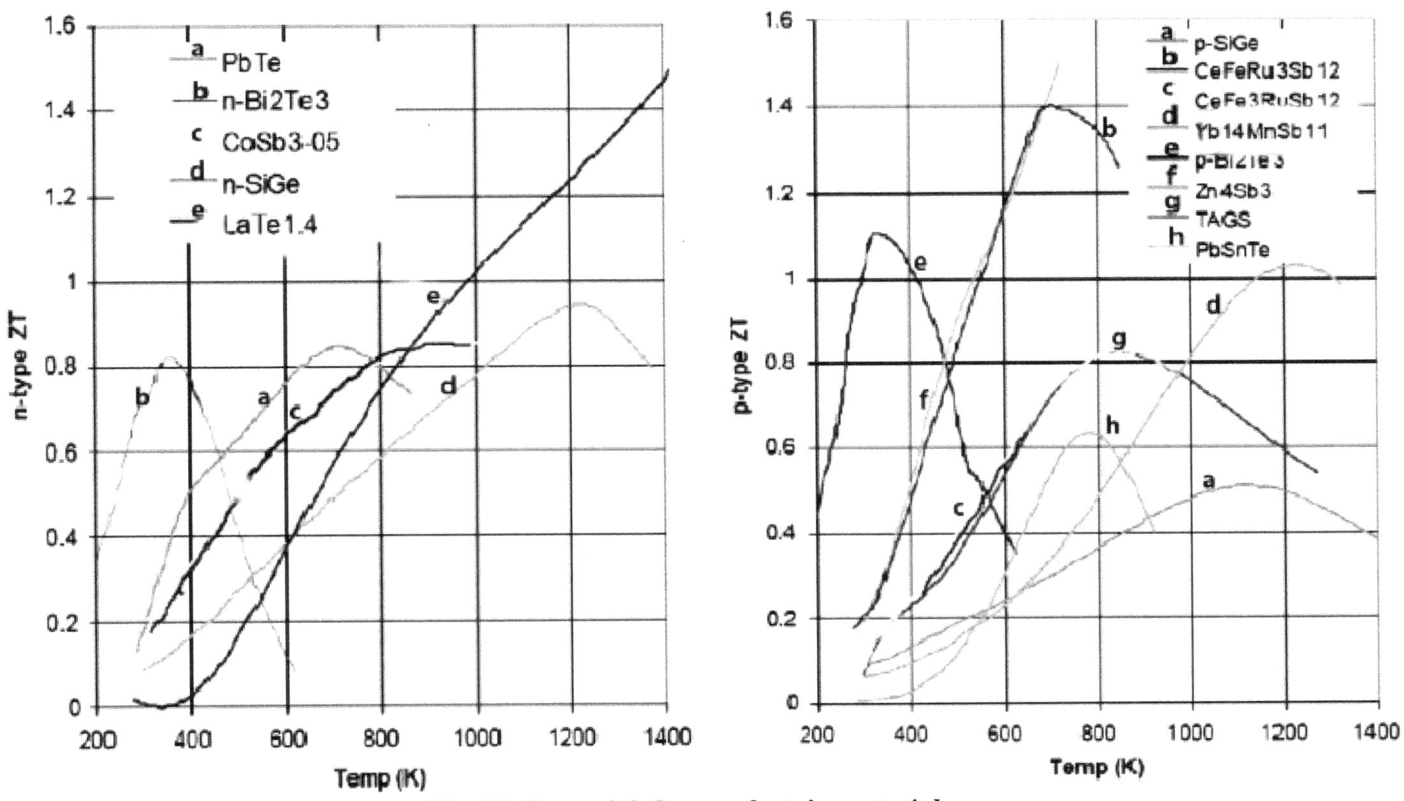

Fig. 27. Potential thermoelectric materials.

Thermal and mechanical properties are also important for applications that need to be stable at high temperatures (near 1275 K) for over 14 years. The sublimation rate must be in the few x 10^{-7} g / cm^2-hr range so that the ARTG does not degrade more than 22% in power output over 14 years. Table 9 provides a comparison of beginning-of-life (BOL) sublimation rates and coefficient of thermal expansion at 1273 K for $La_{3-x}Te_4$, $Si_{0.8}Ge_{0.2}$, and the p-type $Yb_{14}MnSb_{11}$ Zintl phase. For $Si_{0.8}Ge_{0.2}$, satisfactory sublimation rates are achieved with thin coatings of Si_3N_4 / SiO_2.

Thermal expansion matches for the p- and n-type legs greatly facilitates thermocouple integration. High temperature measurements of thermal expansion indicate a good match of $Yb_{14}MnSb_{11}$ and $La_{3-x}Te_4$.

Currently, for most of the selected high ZT materials, fabrication of single phase material with reproducible and stable thermoelectric properties has been achieved. Coatings for sublimation suppression are still being developed.

Table 9. Comparison of beginning of life (BOL) sublimation rates and coefficient of thermal expansion for selected advanced thermoelectric materials.[41]

Material	BOL Sublimation rate ($g/cm^2/hr$)	Coefficient of Thermal Expansion ($10^{-6}/K$)
$Si_{0.8}Ge_{0.2}$	~ 8 x 10^{-3}	5.0
$Si_{0.8}Ge_{0.2}$ (Coated)	~ 5 x 10^{-7}	5.0
$Yb_{14}MnSb_{11}$	~ 7 x 10^{-3} (1)	19.9 (2)
$La_{3-x}Te_4$	~ 5 x 10^{-5}	19.4

(1) Paik, 2007
(2) Ravi, 2007

Alkali Metal Thermal-To-Electric Converter (AMTEC)

AMTEC (Alkali Metal Thermal-To-Electric Converter) is a thermally regenerative electrochemical device for the direct conversion of heat energy to electric power. It is characterized as operating at high potential efficiencies (near Carnot) and no moving parts. The device is a sodium concentration cell which uses a ceramic, polycrystalline beta"-alumina solid electrolyte (BASE) as separator between a high pressure region containing sodium vapor at 900 - 1,300 K and a low pressure region containing a condenser for liquid sodium at 400 - 700 K.[42]

As a power system, AMTEC is expected to deliver ~ 20 W_e / kg with 15 - 20 % thermal-to-electric conversion efficiency.

Principles of Operation[43]

In an AMTEC cell, BASE is used as the separator between high and low pressure sides and the ceramic is coated on both sides with thin film, porous metal electrodes. When sodium is heated on one side of the BASE in a vacuum environment, a vapor pressure difference occurs across the BASE. The pressure difference results in conduction of sodium ions through the BASE. On the high-pressure side of the BASE, an anode is used to collect electrons that dissociate from the neutral sodium and conducts them to a cathode placed on the low-pressure side of the BASE. Here, sodium ions recombine with the electrons to reform neutral sodium. The flow of electrons is the electric current to the external load.

Fig. 28 shows a diagram of the overall system in a recirculating AMTEC cell. Recirculation is accomplished using capillary pumps, or wick systems. Fig. 29 shows a detail of the processes at the interfaces between the electrolyte (BASE) and the electrodes.

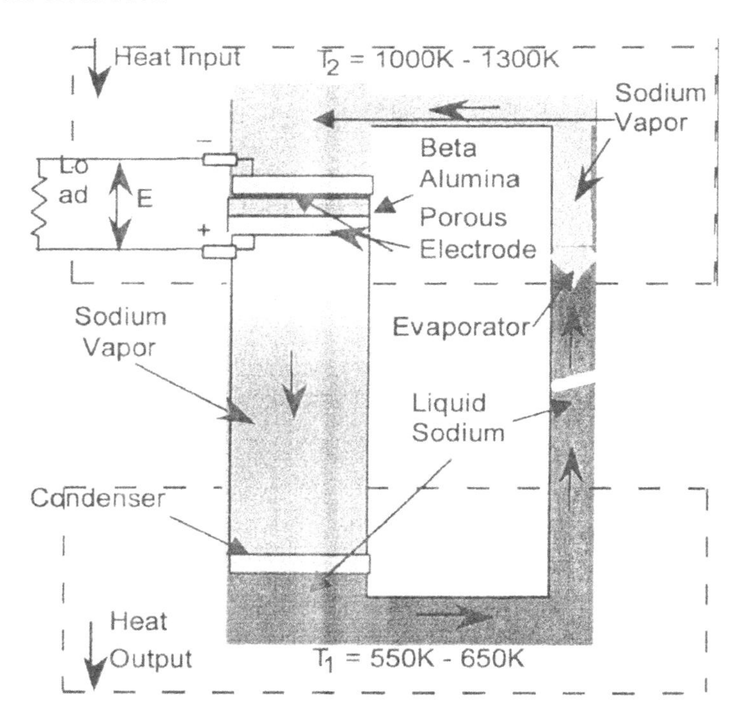

Fig. 28. Schematic of a recirculating AMTEC cell.[44]

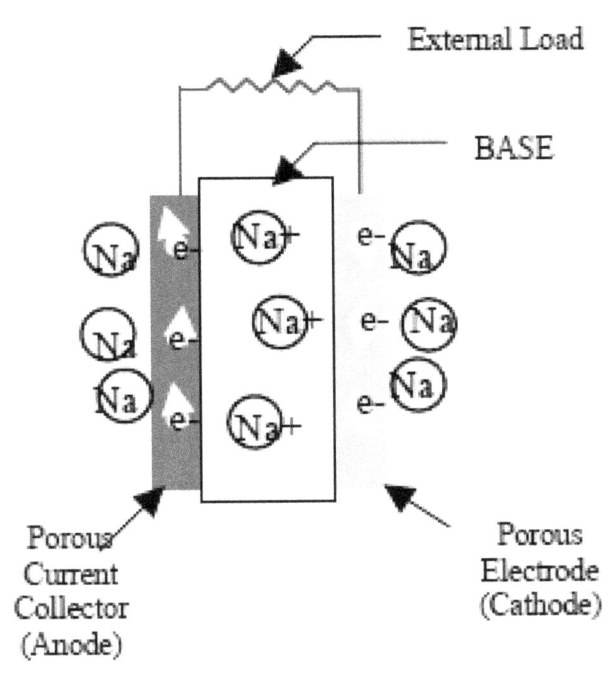

Fig. 29. Illustration of AMTEC conversion process.

In the AMTEC cycle, liquid sodium at T_1 is pumped into the hot zone by the capillary wick where it absorbs thermal energy. By the time it reaches T_2 in the evaporator, it is sodium vapor. On the high pressure side of the BASE, the sodium is ionized. The pressure difference between the anode and cathode sides of the BASE drives Na^+ ions across the separator and electrons travel through the load. At open circuit, Na^+ is driven toward the low pressure side by thermal kinetic energy, and a potential difference is created by charge build-up on the cathode side. Using the Nernst equation, the open circuit potential E_{oc} is:

$$E_{oc} = RT_2/F \ln(P_2/P_3) \quad (13)$$

where R is the gas constant,
T_2 is the hot side temperature,
F is the Faraday constant, 96,485 C / mol.,
P_2 is the vapor pressure of sodium at T_2,
P_3 is the vapor pressure of sodium at the cathode BASE interface.

The vapor pressure of sodium at the cathode BASE interface, P_3, is related to the sodium vapor pressure at T_1, the condenser temperature, as:

$$P_{3(i=0)} = P_1(T_2/T_1)^{1/2} \quad (14)$$

High efficiency results from the large entropy change associated with the change in state from liquid to vapor phase sodium and the incompressible nature of the liquid phase. The AMTEC cell can operate near Carnot efficiency; the reversible thermodynamic cycle is inherently efficient and relatively insensitive to losses related to pressurizing and heating of the liquid sodium and cooling and condensing the low pressure vapor.

There are several sources of losses in a real AMTEC device. These include electrochemical, thermal, electrical, and degradation components over time. The open circuit voltage of a single electrochemical cell can run as high as 1.6 V at a BASE temperature of 1,120 K. Typical open circuit voltage is 1.2 - 1.4 V. Open circuit voltage is impacted by the hermetic seal between the high and low pressure sides and pressure losses caused by "leakage current" or sodium ion conduction accompanied by electronic conduction through the BASE.

The primary electrochemical factor is over-potential--the deviation of the cell potential from its equilibrium value. The over-potential at the anode is insignificant compared to that at the cathode. The over-potential at the cathode is influenced by the exchange current at the electrode / electrolyte interface (a measure of the efficiency with which sodium ions are reduced by electrons entering from the cathode) and by the mass transfer loss coefficient. The mass transfer loss coefficient can be translated into a morphological parameter, G. G is a measure of the resistance to sodium atom flow from the electrode / electrolyte interface to the outside surface of the cathode at the location where it can enter the vapor space. These values are determined experimentally.

Ohmic losses can also be significant. These are primarily of sheet resistance in the cathode, contact resistance between current collecting screens and the cathode, contact resistance between the screen and the lead, and lead resistance.

Leakage between the high and low pressure sides of the BASE also is a source of loss in the current-voltage characteristics of a AMTEC cell. The current-voltage (iV) curve will be depressed by any leak in sodium pressure between the high pressure side and the low pressure sides. Sodium leaks may be caused by non-hermiticity in the seals where the BASE is joined to the cell assembly, and by any other source of "leakage current". "Leakage current" is attributed to electronic conductivity in the BASE and electronic conduction in the braze seals. Non-hermiticity is a much more significant source of loss in AMTEC cells than losses by "leakage current".

Fig. 30 illustrates how current-voltage curves of an AMTEC cell changes when the over-potential of the anode and cathode, ohmic losses in the cell, and open circuit voltage loss are taken into consideration.

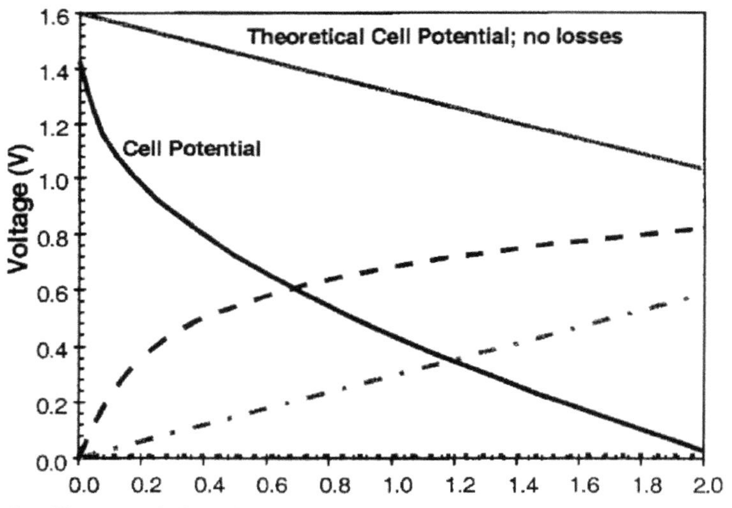

Fig. 30. Theoretical cell potential and actual cell potential showing losses. Dashed line - cathode over-potential, dotted line - anode over-potential, dot-dash - ohmic losses.

Thermal losses consist primarily of radiative losses and some conductive loss. Conductive loss is a thermal loss in the containment of the cell. This leads to a need for a higher condenser temperature than is optimum. Radiative losses can occur in thermal losses from BASE to cell walls.

AMTEC Cell Design

Early designs of an AMTEC cell had difficulties in fabrication. As a result a revised cell design named a "chimney cell" was devised (Fig. 31). In this design, the BASE tubes separates the mounting flange from the hot parts of the cell. The chimneys of the cells are on the end of the general-purpose radioisotope heat source. These are supported with a block of insulation within the cell housing. Performance analysis has shown a 22% heat-to-electricity conversion efficiency, with possible improvements to 24%. Advanced features under development include use of molybdenum-rhenium alloy for cell walls and tungsten-rhodium alloy for electrodes.

Life test data from earlier converter designs operated for over 10,000 hours. This converter used refractory metals Nb-1%Zr. The converter housed eight BASE tube assemblies each with a 1.0 cm diameter tube and an active electrode length of 2.5 cm.[45]

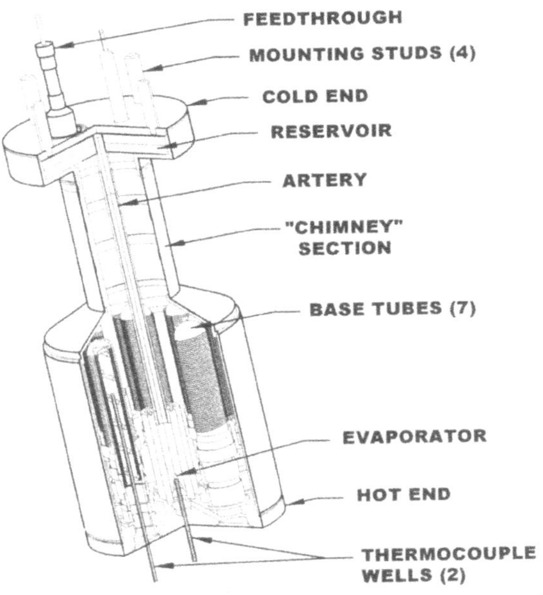

Fig. 31. Chimney converter configuration.[46]

Material stability is a critical issue in AMTEC design. Individual materials must be able to survive as well as the potential deleterious interactions between materials understood. Volatile and reactive elements need to be removed from hot regions of the AMTEC cell where their volatility would be non-negligible over the lifetime of the device. Earlier devices using copper, molybdenum, refractory metals, Haynes 25 alloy, were unsuitable because they failed to take the material stability and interactions into account. Molybdenum/rhenium has been selected as a better material choice because its properties are thoroughly documented and meets AMTEC cell requirements. However, it has a relatively high thermal conductivity from the hot to the cold side and limited temperature range in an oxygen-containing environment.[47]

Similar constraints exists between both the BASE and insulating components high temperature seal and the wick that transports liquid sodium from the low temperature end of the AMTEC converter to the high temperature plenum. The switch to refractory metal wicks still has problems with occasional dry out and leaks, but in general these have been successful. The behavior mechanisms are not well understood and further modeling and testing is needed.

The performance and life characteristics of the rhodium/tungsten electrode are such that a satisfactory electrode design is thought to exist.

Sodium beta"-alumina solid electrolyte electrodes issues can be considered to be solved in principle, though quality control issues remain. Studies indicate the kinetic stability of sodium beta"-alumina to 1,273 K in vacuum or sodium vapor, and to 1,173 K in liquid sodium for hundreds to thousands of hours. At 1,273 K or above, mass losses occur of sodium carbonate, Na_2CO_3, or sodium meta aluminate, Na_2AlO_2, from BASE.

Both phases can be eliminate from BASE ceramic by annealing it in an air atmosphere under a gently packed mixture of powdered sodium beta"-alumina (80%) and sodium beta-alumina (20%) for 100 hours at 1,673 K.

Optimum AMTEC performance depends on low internal cell impedance, while preventing internal discharge modes. Low internal impedance has been demonstrated, but not at the device voltage which represents realistic operating conditions for AMTEC. The design uses eight BASE tubes in a common low pressure chamber and connected to a common high pressure sodium vapor plenum. All but one of these tubes are mounted on insulating stand-offs composed of alpha-alumina or sapphire. However, testing has shown that sodium reactions with alpha alumina or sapphire are possible, but slow. Additional insulating materials research is needed.

Plasma-short through the alkali metal vapor between electrodes have been noted at voltages greater than 5.5 volts. The plasma shorting may be contributing to losses in some Advanced Modular Power Systems, Inc. (AMPS's) series connected cells.

To achieve optimum AMTEC performance, it is necessary to maintain as large a sodium activity gradient across the BASE ceramic while requiring efficient transport of sodium to the condenser. Parasitic heat flow to the cold end must also be limited. Refractory wicks in the design appear to be a successful development. More work is needed in understanding how these wicks functions.

A large amount of power generated is lost due to the low open circuit voltage, which is typically about 0.5 V in current multi-tube converters. In multi-tube series connected arrays of a typical size, each cell should be producing an open circuit voltage of 1.1 volts. Measurements in such arrays show the maximum voltage to be round 0.5 - 0.6 volts per cell. The mechanisms must be understood to obtain adequate performance and life.

A representative AMTEC design capable of providing 141 W_e BOL is shown in Fig. 32.[48] A total of sixteen AMTEC cells are arranged radially around four GPHS modules. The converter hot side temperature is 1,123 K and cold side 655 K. Power conversion efficiency is 15.4%. The total system mass is 20.67 kg to give a specific power of 6.8 W_e / kg BOM.

AMTEC is considered to be at a technology level where analytical and experimental critical function and/or characteristic proof-of-concept have been achieved (NASA Technical Readiness Level of 3 0).[49] For a beginning-of-mission power level of 120 watts-electric, the system mass would be 13.6 kg. This gives a specific power of 8.8 We/kg, about twice that of a SiGe thermoelectric RTG. System efficiency is 16.7%. Life issues are seals, wick, evaporator, containment, materials, and fabrication process. Spacecraft interface issues still to be resolved concern launch vehicle acceleration. Resiliency to partial failure would result in partial power loss.

AMTEC Summary

AMTEC has the potential of achieving system efficiencies of 20% and doubling the specific power over current SiGe RTG systems to 9 W_e / kg. However, major challenges exist:[50]

- Development of a BASE to metal ceramic seal,
- Development of a converter refractory metal containment material fabrication process,
- Development of a reproducible wick-evaporator fabrication process, and
- Development of an electrical feed-through fabrication process.

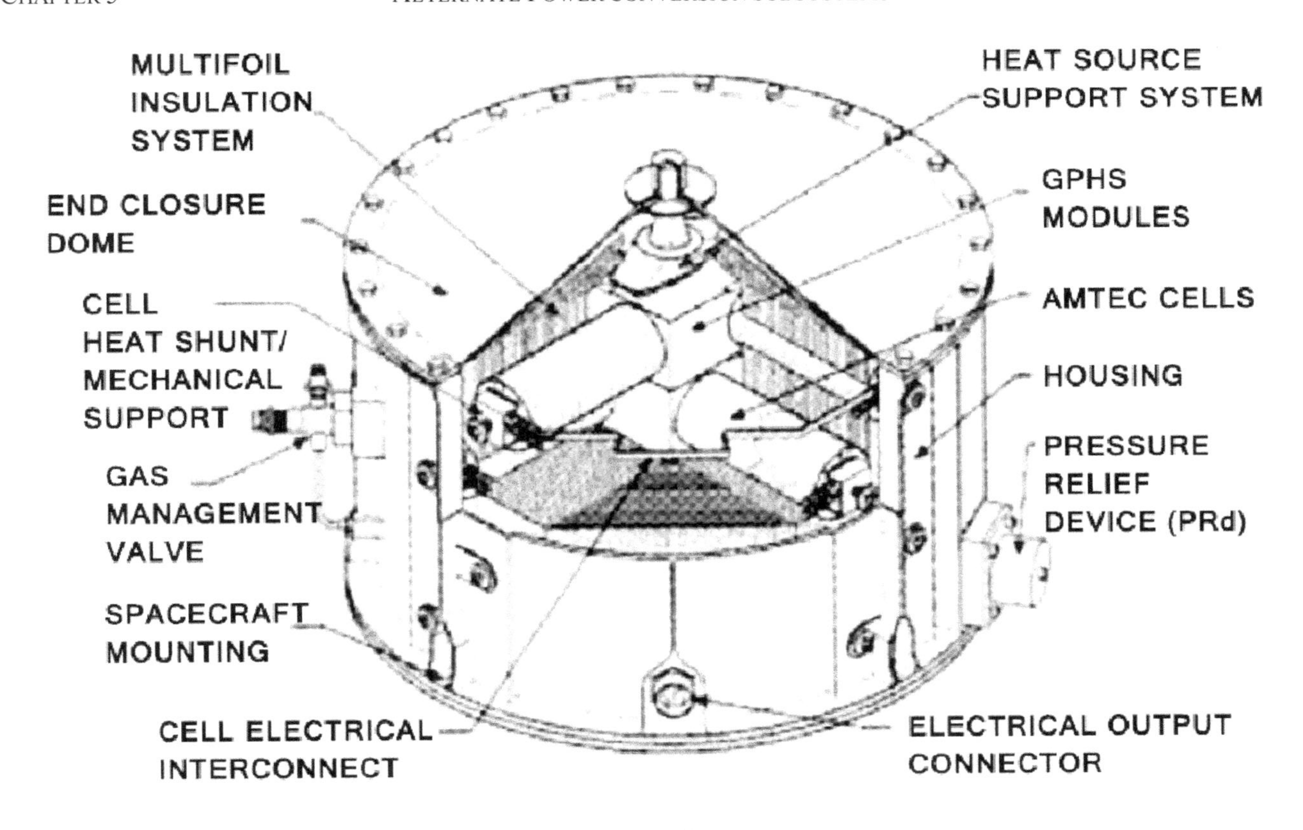

Fig. 32. AMTEC generator.

Thermophotovoltaic (TPV) Generators

Thermophotovoltaic converters for use with radioisotope generators are at a research level. It is an offshoot of photovoltaics in which infrared photons from a heated emitter are converted to electricity. While solar radiation comes from a heat source (Sun) operating at about 3,273 K, radioisotope heat sources typically operate in the 1,273 K to 1,473 K range.[51] New advances in photovoltaics and compound semiconductor wafers have substantially increased cell performance and has let to an interest as a viable passive energy conversion technology. A significant advantage of TPV generators is an inherent redundancy in the design and much higher converter efficiency then thermoelectric units.

The thermophotovoltaic process, illustrated in Fig. 33, converts the radiant heat from the GPHS heat source to electricity by adding a hot emitter surface to the heat source. A filter component provides spectral control to facilitate transmission of radiation in shorter wavelengths usable by the TPV cells and reflects radiation in longer wavelengths unusable by the TPV cells. The TPV cell converts a portion of the transmitted radiation to electricity. The reminder of the transmitted radiation is rejected to a heat sink, to radiators on the spacecraft.[52] TPV cell conversion efficiency is strongly dependent on both emitter and cell temperatures. Fig. 34 and 35 show these impacts for various materials and temperatures.[53]

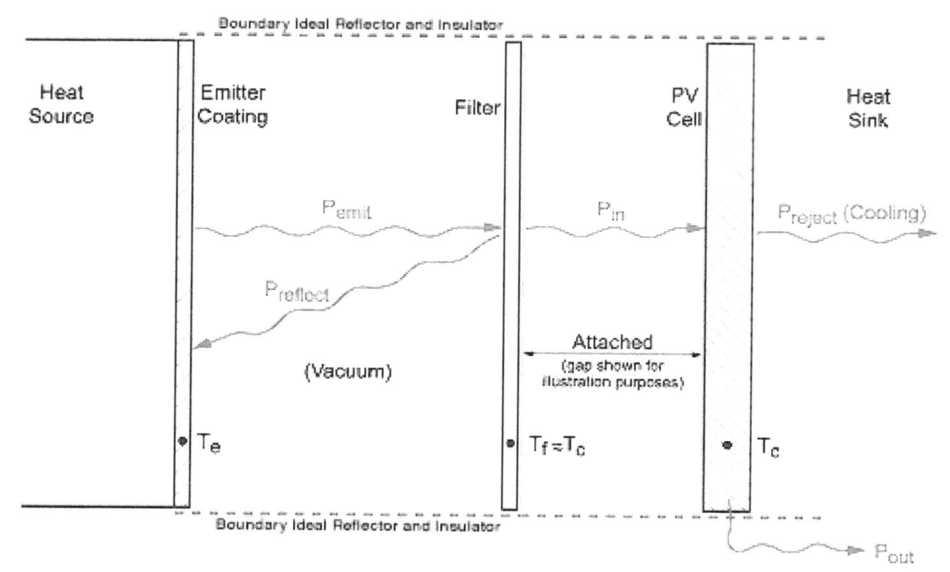

Fig. 33. Thermophotovoltaic power conversion process.

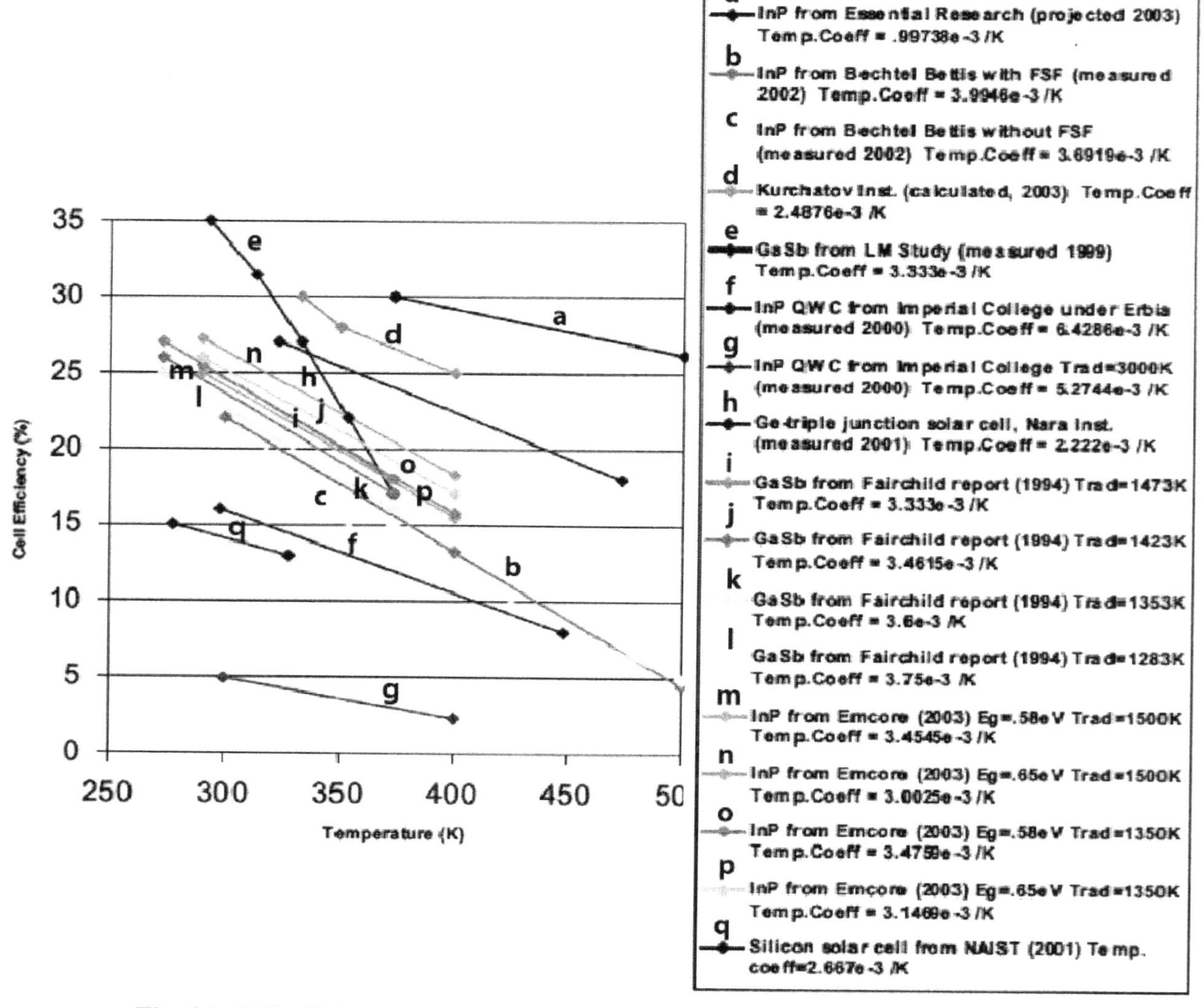

Fig. 34. Cell efficiency vs cell temperature for various possible cell materials.

An example of a 500 W_t system TPV design is shown in Fig. 36.[54] Two GPHS units are installed in a cube approximately 10 cm on a side and supported in a canister. The canister is in turn enclosed by an aluminum housing assembly. Multifoil insulation in the housing minimizes parasitic heat losses. The TPV converters are located on each of the two end faces of the housing. The temperature range is limited to between 1,200 K and 1,350 K to keep the iridium cladding on the fuel pellets within acceptable limits. The heat flux on each TPV array for the two faces of 100 cm^2 is about 2.5 W / cm^2.

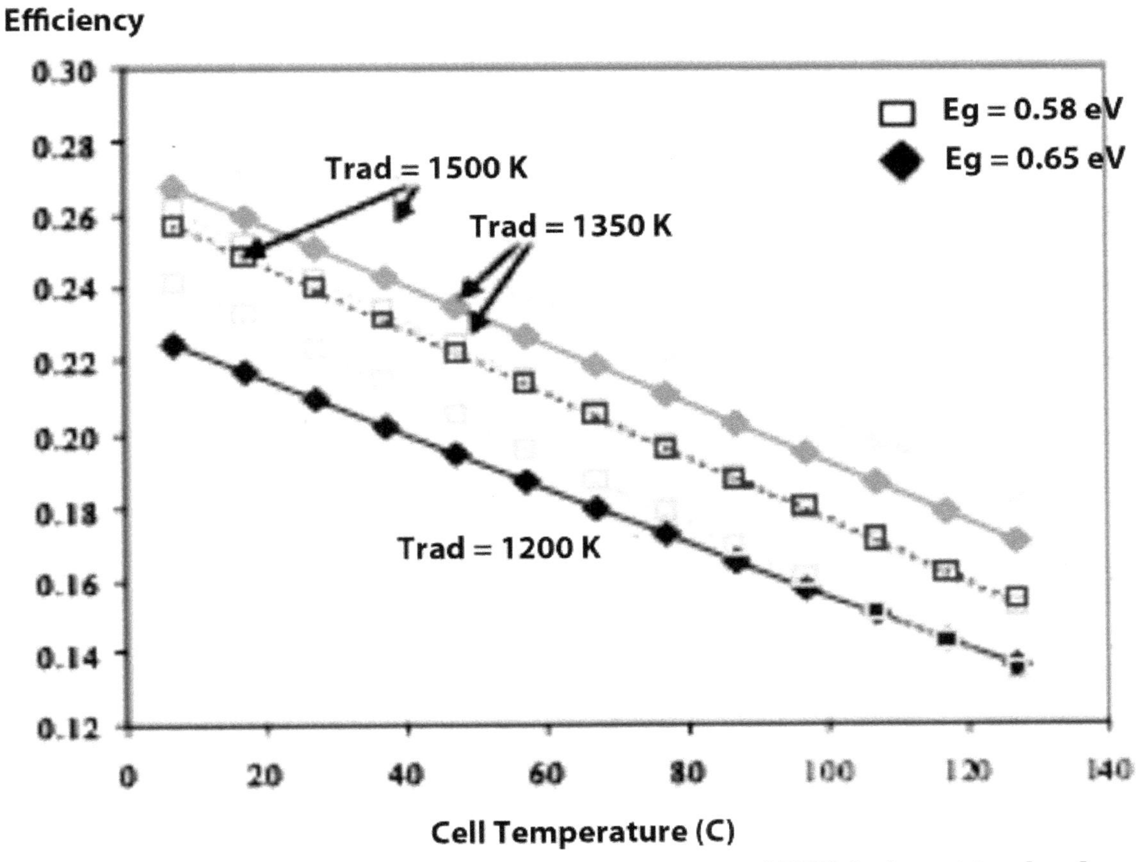

Fig. 35. Cell efficiency versus cell and radiator temperature of TPV devices at two bandgaps.

Several materials are available for the TPV cells. In this array InGaAs on InP substrates was used. The dielectric filter is comprised of multiple layers of high index of refraction material (Sb_2Se_3) and low index of refractor material (YF_2). The plasma layer is InPAs on an InP substrate. The filter has very low reflectance (~ 0.1 microns) convertible to electric power by the TPV cells, and very high reflectance in the range of wavelength (> 2.07 microns) not convertible by the TPV cells. The emitter is a graphite plate.

Using the above TPV design, measured cell and cell arrays and analytical estimates extending these measurements to a complete generator, performance for a 500 W_t RPS are given in Table 10. The projected specific power for such a unit is 12 W_e / kg, with future growth to 14 W_e / kg.

In summary, thermophotovoltaic generators show promise as a high efficient, redundant systems, but need much more effort to become a viable contender for future missions.

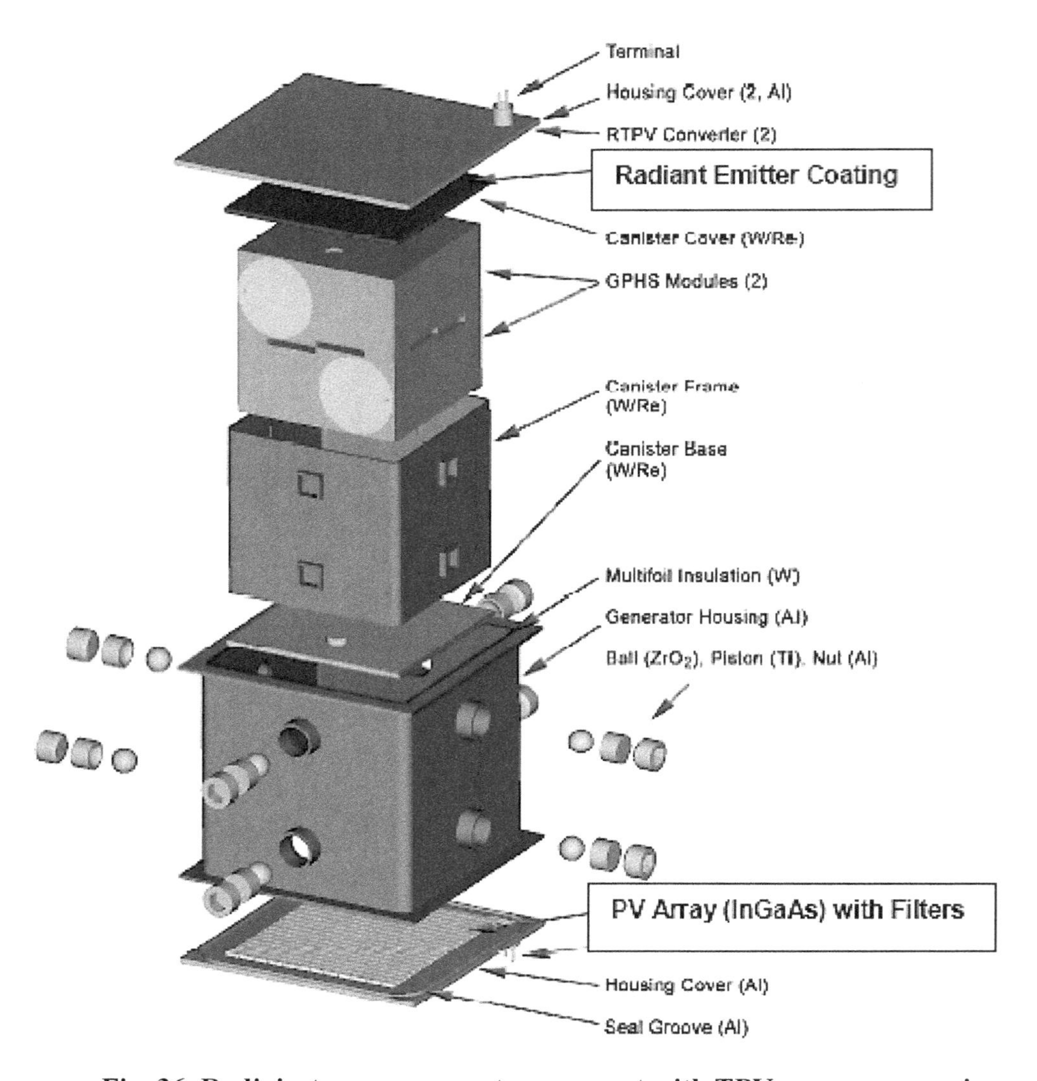

Fig. 36. Radioisotope power system concept with TPV power conversion.

Table 10. Estimates of TPV performance for a 500 W_t radioisotope power system.

Generator Level	Efficiency Factor	Current Basis	Current Value	Future Value
Cells only	η_{PVcell}	Measured Data	0.28	0.32
Filtered Cells	$\eta_{spectral}$	Measured Data	0.8	0.8
Cell Arrays	η_{cavity}	Measured Data	0.9	0.9
Cell Arrays	$\eta_{network}$	Measured Data	0.98	0.98
Generator	$\eta_{housing}$	Estimated Analytically	0.9	0.9
Generator	$\eta_{DC\ Control}$	Estimated Analytically	0.98	0.98
Total	η_{RTPV}	**Estimated Analytically**	**0.17**	**0.20**
	Output Power	**Estimated Analytically**	**85 W_e**	**100 W_e**
	Mass	**Estimated Analytically**	**7.1 kg**	**7.1 kg**
	Specific Power	**Estimated Analytically**	**12 W_e/kg**	**14 W_e/kg**

Summary

A number of higher efficiency, alternate power conversion systems have been studied: to extend the operating range of radioisotope power systems to about 10 kW_e; to require the use of smaller quantities of plutonium-238 which is in short supply and expensive to produce; and to improve the specific power of the units. We have reviewed the converters that have received the most engineering efforts to date.

Table 11 provides some representative values for the various converter systems discussed. The GPHS-RTG, currently used in recent RTG missions, provides a reference system. The system efficiency relates directly to the quantity of plutonium-238 required for a given power level. Thus, high efficiency lowers the plutonium-238 required. The specific power is a figure-of-merit for system mass. Significant improvements are highly desirable. The converter hot side temperature establishes the materials needed in the radioisotope unit and components of the converter. The current GPHS modules have been qualified for 1,300 K operation. The radiator inlet temperature, since heat rejection area to space is proportional to the forth power of the radiator temperature, establishes the radiator size.

NASA uses Technology Readiness Levels (TRLs) as a systematic metric/measurement system that supports assessments of the maturity of a particular technology and consistent comparison of maturity between different types of technology.[55] These levels are summarized below.

TRL 1 Basic principles observed and reported

TRL 2 Technology concept and/or application formulated

TRL 3 Analytical and experimental critical function and/or characteristic proof-of-concept

TRL 4 Component and/or breadboard validation in laboratory environment

TRL 5 Component and/or breadboard validation in relevant environment

TRL 6 System/subsystem model or prototype demonstration in a relevant environment (ground or space)

TRL 7 System prototype demonstration in a space environment

TRL 8 Actual system completed and "flight qualified" through test and demonstration (ground or space)

TRL 9 Actual system "flight proven" through successful mission operations

The technology readiness levels represents the current state-of-technology for the various converters. Most mission planers would like the technology to be at least to level 6 and feel much more comfortable at level 9 or as close to it as possible.

Table 11. Comparison of high efficiency converter performance to GPHS-RTG.

Converter Technology	System Efficiency	Specific Power We/kg	Converter Hot Side Temperature, K	Radiator Inlet Temperature, K	Technology Level
GPHS-RTG*	6.6	5.4	1275	566	9
Rankine	18	6	630	375	5 - 6
Brayton	23	6 (2 - 3.1 with redundant converters)	1130	401	5 - 6
Stirling ASRG-650 C ASRG-850 C	25	7 > 8.5	1125	363	6 3-4
Advanced Thermoelectric Generator	8 - 10	6 - 8	to 1275		3
AMTEC	16.7	8.8	900 - 1300	400 - 700	3
Thermophoto-voltaic Genertors	17 - 20	12 - 14	1200 - 1350		3

Chapter 6

Radioisotope Heater Units

Radioisotope heat source technology developed for electric power generation is also applicable to spacecraft, surface science packages, and rovers to satisfy thermal control requirements. "Heat-only" applications involve: (1) the thermal control of isolated components by small (typically one watt-thermal) distributed nuclear sources; and (2) providing thermal power directly to cryogenic cooler units for active payload temperature control. An earlier heater design used on the Pioneer spacecraft probe to the outer planets, Jupiter, Saturn and beyond is shown in Fig. 1 and its characteristics summarized in Table 1. Twelve of these one thermal-watt units were used on the Pioneer spacecraft.

For isolated component thermal control, the sources used are small, operate at low temperatures, and pose few engineering problems. On the other hand, the radioisotope heat sources used to power cryogenic coolers are relatively large (typically kilowatt-thermal class) and operate at temperatures comparable to PbTe RTGs. These units require the same kind of handling precautions, emergency cooling, auxiliary cooling, insulation, etc., as found in large radioisotope electric power generation systems. Even with these constraints, however, the overall size and weight of such heat sources are extremely favorable when compared to other spacecraft thermal control design options.

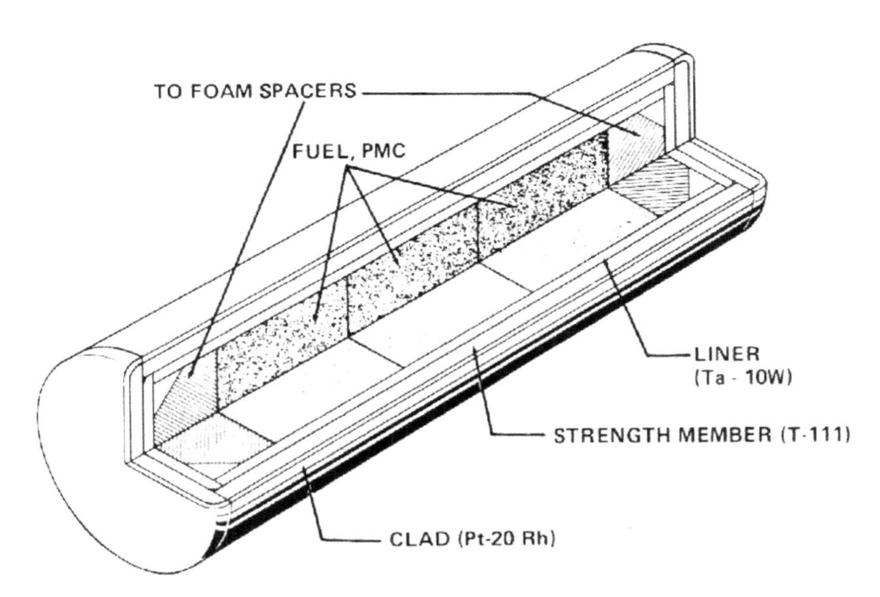

Fig. 1 Pioneer radioisotope heater unit. ***Courtesy of Mound Laboratory Monsanto Research Corporation.***

Over 240 Radioisotope Heater Units (RHUs) have been used in space missions to maintain structures, systems, and instruments sufficiently warm to operate effectively. These RHUs are highly reliable, operate continuously, and provide a predictable output of heat. They have the added advantage of being able to be located right at the point where temperature control is needed. In addition, they have no moving parts, compact, resistant to radiation and meteorite damage, and produce heat independent of distance from the sun.

Table 1. Characteristics of Pioneer radioisotopic heater unit

Fuel Data	
Fuel Form	PMC (Plutonia Molybdenum Cermet)
Fuel Required	
Weight Pu-238/Source (grams)	1.8
Weight Fuel/Source (grams)	3.02
Heat Sources	
Number Fabricated	50
Watts-thermal/Source	1
Operating Temperature (K)	310
Materials of	
Construction Liner	Ta-10W
Strength Member	T-111
Clad	Pt-20 Rh
Geometry (cm)	Cylindrical (2.9 x 1.0)
Weight/Source (grams)	23

RHUs used in space generate heat from the radioactive decay of small pellets of plutonium dioxide. This heat is transferred to spacecraft structures, systems, and instruments directly without moving parts or intervening electronic components. The units are very compact, measuring 3.2 cm long and 2.6 cm in diameter (see Fig. 2). The fuel itself weighs approximately 2.7 grams. Each RHU weighs about 40 grams and produces about 1 watt of thermal power. The brick-like ceramic form of the fuel pellets (a) is designed to break into large pieces, instead of dust, on impact. This makes it difficult to inhale and minimizes its exposure to the environment. In case of an accident, the cladding (b) and insulator (c) protect the fuel pellet from the extreme heat of re-entry and from the environment.[1]

Fig. 3 illustrates the current RHU design. Containment is achieved by use of multiple layering of materials that are resistant to the heat and impact forces that might occur during a spacecraft accident. This includes a reentry shield made of a graphite aeroshell and a graphite insulator to protect the fuel from impacts, fires, and atmospheric reentry conditions. The fuel itself is encapsulated in a high-strength, platinum-rhodium metal shell or clad that further contains and protects the fuel during any potential accidents.

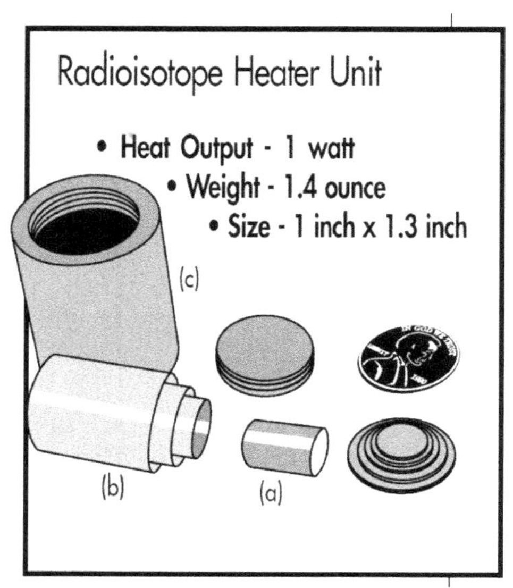

Fig. 2. Radioisotope heater unit. ***Courtesy of U.S. Department of Energy.***

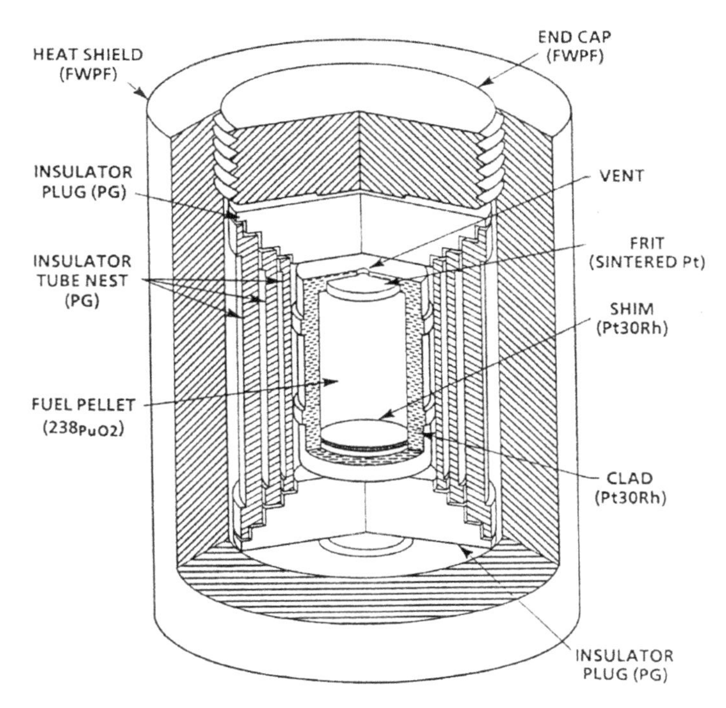

Fig. 3. Lightweight radioisotope heater unit. ***Courtesy of Department of Energy***

In the extremely unlikely event that the containment barriers are breached, the fuel is in a ceramic form. This material tends to break into large pieces rather than dispensing as fine particles. Large pieces further reduce the potential for human exposure.

Safety testing has been used to verify the RHUs design. These tests demonstrated that no radioactive material releases occurred for the range of anticipated accident scenarios.

Radioisotope Heater Unit Design[2]

RHUs are designed to be very rugged even if subjected to severe accident conditions. This is to prevent or minimize the release of plutonium dioxide fuel. Current design requirements for the Light-Weight Radioisotope Heater Unit (LWRHU) are:

- Thermal power of 1.1 ± 0.03 W (beginning-of-life),
- Useful life of 7 y (minimum),
- Neutron emission < 6,000 n / s-g $^{238}PuO_2$,
- Mass ≤ 40 g,
- Surface temperature of graphite aeroshell in free air ≤ 373 K,
- Long-term thermodynamic compatibility of components,
- Performance unaffected by transport and launch vibration,
- Reentry ablation $\leq 50\%$ of aeroshell thickness,
- No fuel release in hard-surface reentry impacts at velocities ≤ 49 m / s,
- 473 K temperature margin (fuel containment cladding) during reentry, and
- Corrosion resistance (no environmental interaction after reentry and Earth impact).

Similar to RTGs, $^{238}PuO_2$ is the fuel of choice because of its high specific thermal power and inherent safety. Safety features include: emits little penetrating radiation, is unaffected by exposure to air, and has a relatively high melting point (2,523 K). For the LWRHU design, the selected fuel was a single, 83.5% enriched $^{238}PuO_2$, hot-pressed pellet, sintered to 86% of theoretical density (specific thermal power = 0.42 W_t / g). Studies showed that a right circular plutonia cylinder with an L / D of 1.5 could be manufactured with good reproducibility. The fuel mass and density values fixed the pellet dimensions at a diameter of 6.25 mm and a length of 9.37 mm. To prevent chipping and to provide a stand off void for the final encapsulation weld, chamfers were added at both ends. Subsequently, pellets pressed to these dimensions demonstrated an average thermal output of 1.106 W_t and an average neutron emission of 5,190 n / s-g $^{238}PuO_2$.

The fuel containment design requires a high melting or eutectic point (at least 200 K above the maximum temperature experienced during reentry), sufficient strength and ductility to survive Earth impact with no fuel release, and chemically compatible with the plutonia fuel, aeroshell, and insulation at predicted reentry temperatures. The platinum-group of alloys tend to meet these criteria with the alloy Pt-30%Rh selected for the containment material. Fig. 4 provides the dimensions for the fuel containment design.

For the fuel vent, a design used in the Multi-Hundred Watt RTG meets the design needs. This included a sintered platinum frit as the filter element (See Fig. 5). Fig. 6 is an exploded view of the fuel capsule.

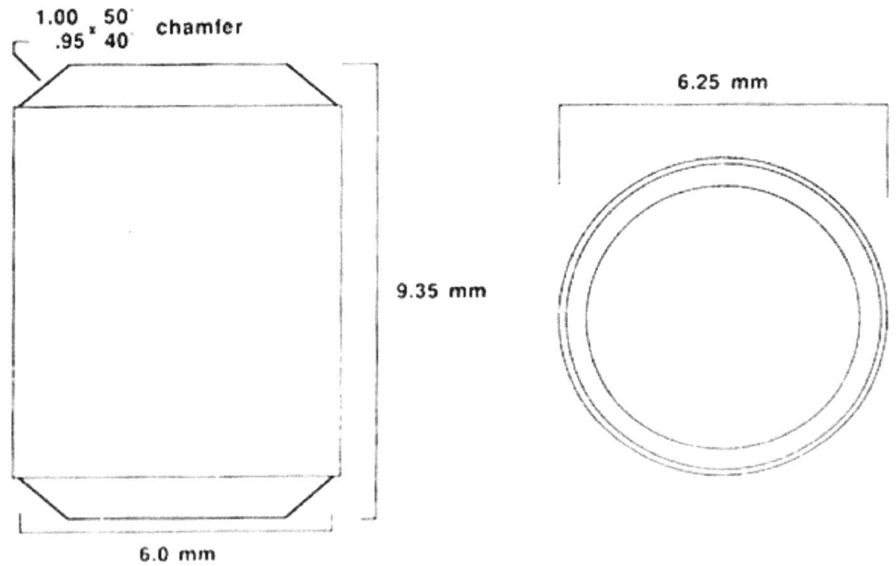

Fig. 4. Dimensions of the LWRHU pellet

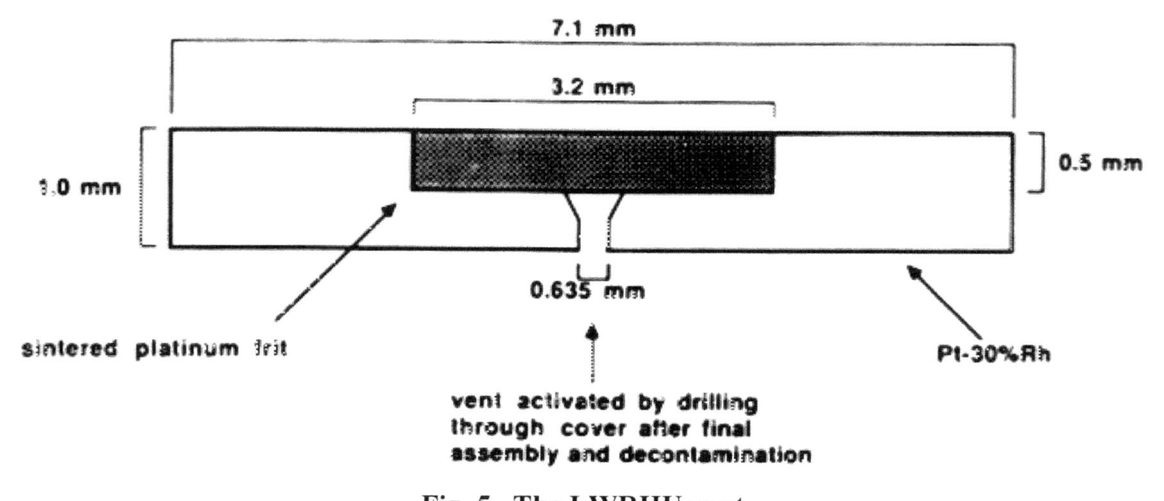

Fig. 5. The LWRHU vent.

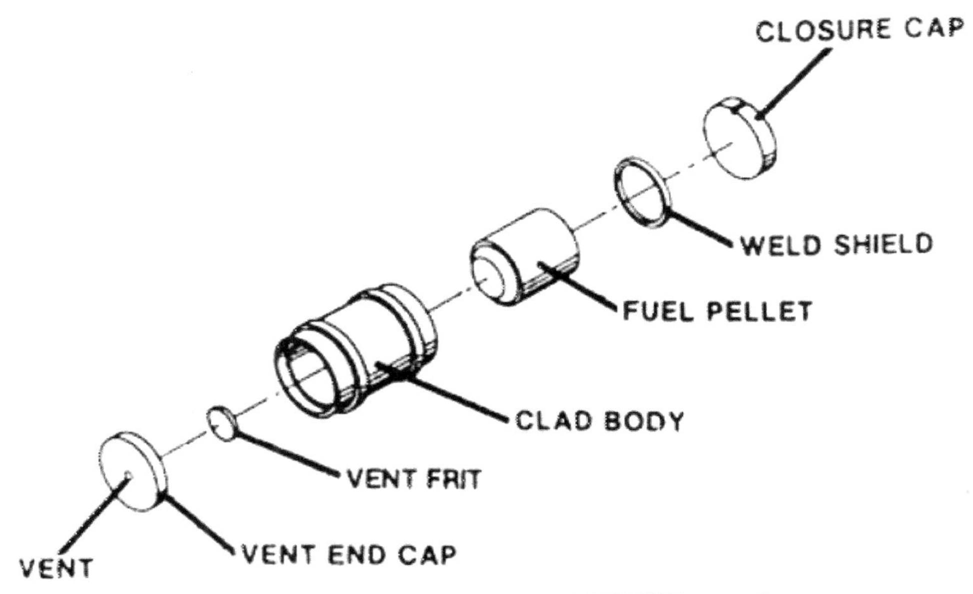

Fig. 6. Exploded view of the LWRHU capsule.

The aeroshell and insulator assembly is shown in Fig. 7. For the insulating sleeve material, pyrolytic graphite was selected based on its strength and its unique thermal properties. Thermal conductivity in the A-B crystallographic plane is 50 to 100 times greater than in the direction perpendicular to the plane. The LWRHU design takes advantage of this anisotropy; reentry heat is conducted around the insulator surface while through-heating (toward the fuel capsule) is retarded. The pyrolytic graphite can only be formed in thin layers. Therefore, three nesting tubes and two end caps are used in the design. Standoff rings were added to the tubes to minimize heat transfer.

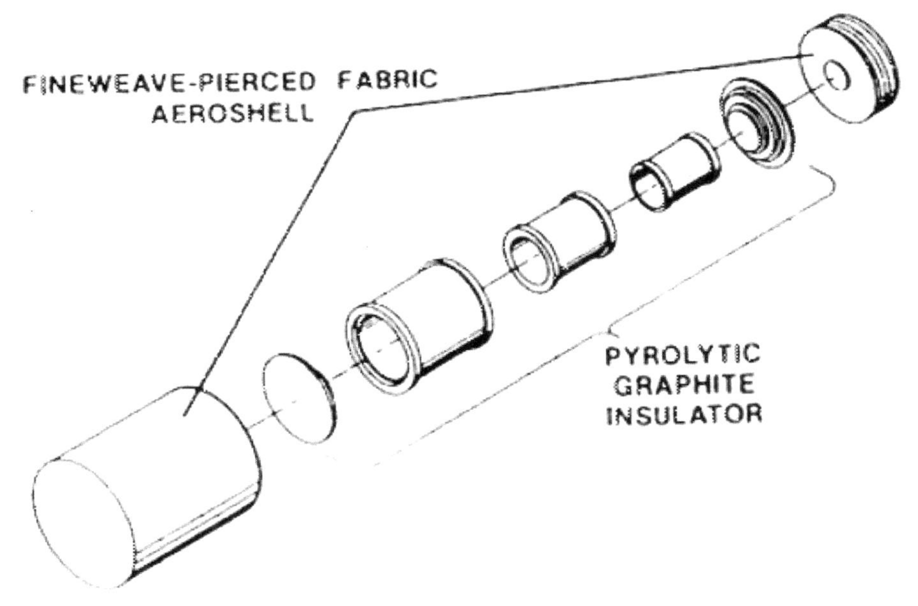

Fig. 7. LWRHU Aeroshell and Insulator Assembly.

The aeroshell uses a carbon-carbon composite of Fine-Weave Pierced Fabric based on testing for the General Purpose Heat Source program. The material has excellent thermal, mechanical, and impact properties; the material retains its strength at very high temperatures, is resistant to thermal shock, does not fail by brittle fracture, and ablates evenly at all orientations. Also, the material is permeable to helium--a consideration during reentry when significant quantities of helium are released from storage in the plutonia. The final design is a thick wall cylinder with a threaded end plug.

Fuel Processing and Fabrication [3]

A fuel processing flow diagram is shown in Fig. 8. For the heater units, a $^{238}PuO_2$ fuel powder results from processing the addition of oxalic acid to plutonium nitrate solution and then calcining the oxalate precipitate at 1,023 K to form plutonia. The feed is then enriched to 80.1 at.% ^{238}Pu and a thermal power of 0.40 W_t / g PuO_2. The initial processing step involved heating the feed powder in an $H_2{}^{16}O$ environment to replace the ^{17}O and ^{18}O isotopes present in the feed with ^{16}O. The original feed powder had an average neutron emission rate of 15,010 n / s-g ^{238}Pu, primarily due to (α,n) reactions causes by the presence of ^{17}O and ^{18}O isotopes in the normal oxygen used in the calcining process. The exchange was accomplished by heating the feed powder in a platinum boat in a horizontal tube furnace in an atmosphere of flowing argon saturated with $H_2{}^{16}O$ for 15 h at 1,048 K followed by 1 h at 1,273 K to release stored helium. After the oxygen-16 exchange and other processing, the average neutron emission rate of the fuel granules is reduced to 4,820 n / s-g ^{238}Pu.

To reduce the differences in surface activity among feed powder lots, the feed powder is ball-milled for 8 h. The average median diameter of about 4 um is reduced to 1 um. The feed powder is then converted into granules by cold pressing 25-g charges at 400 MPa; then breaking up and screening the material to collect the < 210 um granules. This forms a green pellet at 60% of theoretical density. The 60% of theoretical density granules are then sintered for 6 h at 1,373 K in flowing $Ar-H_2{}^{16}O$, while the 40% granules are sintered for 6 h at 1,873 K.

The pellets are fabricated using a hot press system containing a hydraulic system, a vacuum system and an induction heating system. The hydraulic system is capable of producing a maximum force of 111.9 kN. The heating system consisted of a 100 kVA motor generator and associated controls. The fuel to fabricate the fuel pellets blends the < 210 micron granules sintered at 1,373 K (60 wt%) and 1,873 K (40 wt%). The hot press graphite die is loaded with 16 charges of blended fuel, each weighing about 2.675 g. The graphite die is then positioned within the center of the vacuum chamber of the hot press and system evacuated to a pressure of < 6.7×10^{-3} Pa. The pressing is performed under full load for 30 minutes at a temperature of 1,803 K and a force of 11.8 kN. At the end of the pressing cycle, the motor generator is turned off, but the full load maintained on the die assembly for an hour. The load is then removed, the vacuum system backfilled with argon, the die assembly removed, and the pellets ejected from the die.

The hot pressing process reduced the plutonia stoichiometry to $PuO_{1.93}$. To oxidize the pellets back to PuO_2 and increase the density, the pellets are sintered in flowing $Ar-H_2{}^{16}O$ for 6 h at 1,273 K, followed by 6 h at 1,800 K. Heating and cooling rates are 150 K / h. A platinum boat filled with high-fired thoria powder is used to prevent stress points that could occur if the pellets are in direct contact with the platinum boat. After sintering, a typical pellet mass is 2.667 g, a length of 9.34 mm, and a diameter of 6.23 mm.

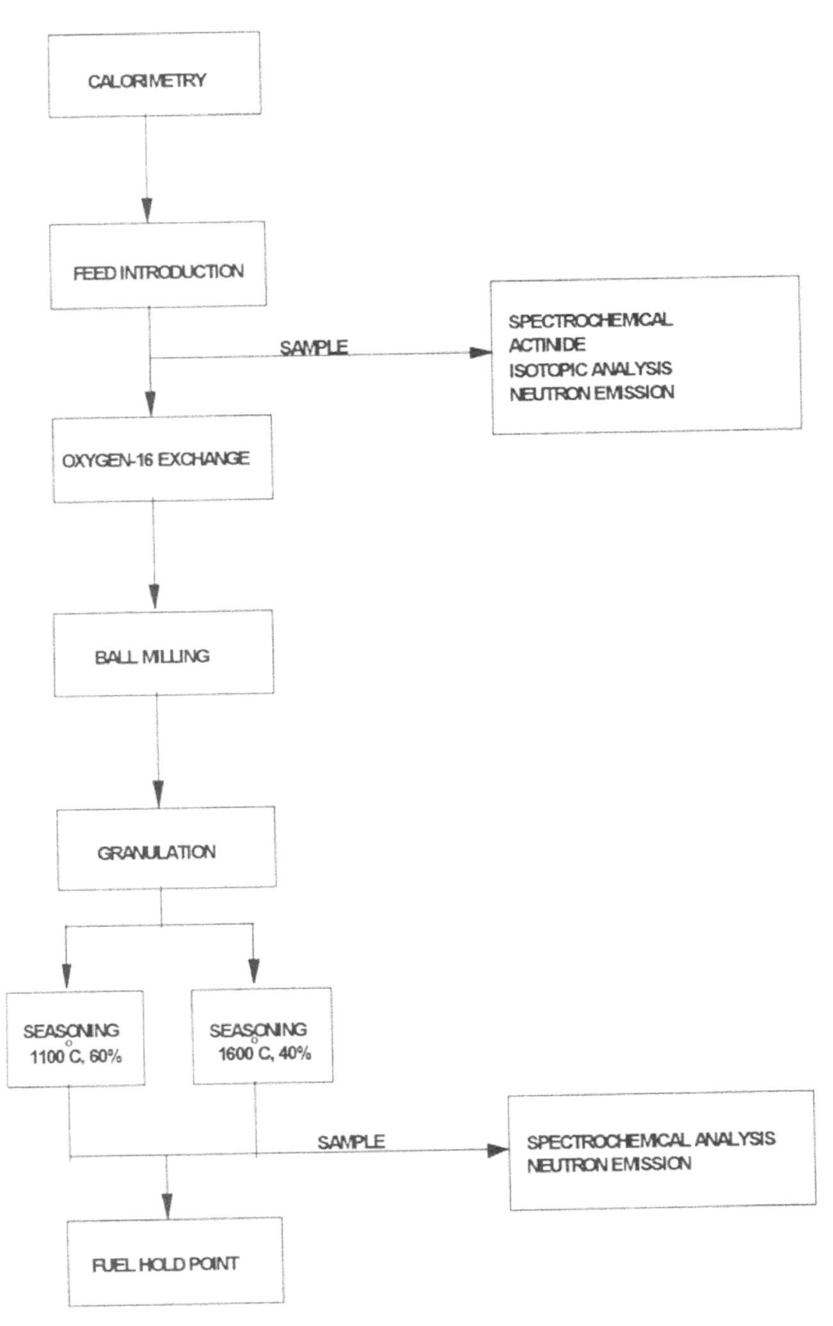

Fig. 8. LWRHU fuel processing flow sheet.

Welding, Nondestructive Testing, And Final Assembly

The capsule components, except for the frit vent, use Pt-30%Rh tube, sheet, and foil. The frit vent is a pressed and sintered disk of pure platinum powder. The frit vent and vent end cap are electron beam welded together. Other components are welded using a gas tungsten arc welding system in a helium atmosphere glove box. A gas purifying system plumbed to the glove box is a closed loop system to maintain the atmosphere purity at < 100 ppm oxygen and < 250 ppm moisture. The glove box is maintained at a pressure approximately 1-2 cm of water negative with respect to the room atmospheric pressure using helium as needed.

The welded fueled clads are tested for helium leaks, radiography of the weld area to ensure 100% weld penetration, measurement of the height, diameter, and weld standoff height, neutron emission rate, and thermal power. Also, the gamma and neutron dose of each fueled clad is measured at 20 cm. The thermal power of each fueled clad is measured using calorimeters.

After each fueled clad passes the nondestructive testing, final assembly is performed. The pressed and sintered platinum powder frit vent is activated by cutting with a milling machine into the vented end cap of the fueled clad at the center line with a 0.635 ± 0.050 mm diameter end mill to a depth of 0.279 to 0.406 mm. The fueled clad is then placed into the graphite aeroshell and the graphite lid glued in place with cement. The graphite aeroshell is then heat treated at 373 K for 4 h followed by heating at 403 K for 16 h to cure the cement. Finally, to ensure that volatile components of the glue do not outgas in the spacecraft, each aeroshell assembly is outgassed in a vacuum furnace at a temperature of 598 K for 144 h at a pressure less then or equal to 2×10^{-5} Torr.

Safety Verification Tests

Safety verification tests to evaluate the response of the RHU to credible accidents are categorized: (1) according to the effects from exposure to environments that would result from an orbital abort of the spacecraft; and (2) the effects of launch accidents. Environmental exposure tests include: reentry simulation, helium release, Earth impact, and seawater immersion. Launch accident tests include: over-pressure, burning propellant, bullet/fragment, and impact tests. The results of safety tests are given in Tables 2 and 3.

Reentry simulation of an inadvertent reentry event could not be totally simulated by laboratory testing. Therefore, a combination of laboratory testing and computer modeling and analysis was used. The analyses indicate that the peak clad temperature during a worst-case reentry would not exceed 1,773 K--which is within the design goal of 200 K below the Pt-30%Rh-C eutectic temperature of 1,973 K. Analysis of whether the rapid release of helium, previously trapped within the fuel, would degrade the thermal effectiveness of the insulator gaps and raise the peak clad temperature beyond an acceptable limit showed only a 40 K additional temperature rise. The performance of the aeroshell for ablation showed that the 50% recession goal was met for the worst-case reentry trajectory and enough aeroshell material remained to prevent release of the capsule during the reentry heat pulse. Laboratory tests were conducted on helium release, Earth impact, and seawater immersion. The result, see Table 2, indicated safety criteria were met.

Helium release tests were conducted to determine if the vent flow rate would be adequate to prevent fuel capsule pressurization and resultant deformation or rupture from helium release during reentry. Two fueled capsules were placed in an evacuated tungsten-mesh resistance furnace. Using the most severe temperature ramp predicted by analysis of raising the temperature to 1,873 K in 90 s, the tests showed neither of the capsules deformed.

Analysis of Earth impact indicated that the impact velocity at sea level would be 44.5 m / s, and its temperature would be no greater than 533 K. Results of the impact tests using a 178-mm-bore gas gun under 'Reentry Impact' indicates minimal capsule deformation. Each test resulted in the graphite components being significantly damaged, but little or no damage or distortion to the fueled capsule. Two units were aged for 2.5 y before testing for impact; these units also did not release any fuel.

Table 2. LWRHU safety verification tests.

Test	Test Condition	Results
Reentry Impact	49.3 m/s, 47 C	Minimal capsule deformation
	49.3 m/s, 30 C	Minimal capsule deformation
	49.0 m/s, 51 C	Minimal capsule deformation
	49.0 m/s, 52 C	Minimal capsule deformation
	49.9 m/s, 24 C, Aged 2.5 y	Cracked weld, not breached
	49.2 m/s, 26 C, Aged 2.5 y	Minimal capsule deformation
Seawater Immersion	2.5×10^{-3} MPa[a], 10 C, 1.75 y	32 μg Pu recovered from graphic ashes.
		2.5 ng Pu/L seawater (max.) detected in weekly sample.
	68.9 MPa[b], 10 C, 1.75 y	154 μg Pu recovered from graphic ashes.
		10.0 ng Pu/L seawater (max.) detected in weekly sample.
Explosion	12.76 $\pm$ 0.69 MPa Overpressure	Graphite stripped from capsule
	38.6 $\pm$ 3.4 MPa-s Impulse	Capsule intact on x-ray, not recovered
	12.76 $\pm$ 0.69 MPa Overpressure	Graphite stripped; capsule integral.
	38.6 $\pm$ 3.4 MPa-s Impulse	One capsule breached (faulty weld). One capsule not recovered.
Solid Propellant Fire	289 m/s	Capsule deformed, not breached
	661 m/s	Capsule deformed, not breached
	773 m/s	Capsule not hit
	775 m/s	Capsule not recovered
	757 m/s	Capsule deformed, not breached
	940 m/s	Capsule fragmented
	908 m/s	Capsule fragmented
Terrestrial Impact	Unvented capsule, side-on, 48 m/s	Capsule deformed, not breached
	Unvented capsule, end-on, 48 m/s	Capsule deformed, not breached
	Unvented capsule, 45°, 48 m/s	Capsule deformed, not breached
	Unvented capsule, side-on, 68 m/s	Capsule deformed, not breached
	Unvented capsule, side-on, 88 m/s	Capsule deformed, not breached
	Unvented capsule, side-on, 105 m/s	Capsule deformed, not breached
	Unvented capsule, side-on, 128 m/s	Capsule deformed, not breached
	Unvented capsule in graphite, side-on, 105 m/s	Capsule deformed, not breached
	Vented capsule in graphite, side-on, 48 m/s	Capsule deformed, not breached

[a] A pressure equivalent to submersion at a depth of 0.25 m.
[b] A pressure equivalent to submersion at a depth of 6,000 m.

Table 3. Safety verification tests at Mound and The University of Dayton.[4]

Test	Test Article	Parameters	Result
Low-Level Overpressure Test (Mound)	Simulant-fueled LWRHUs	2.96 MPa 34.5 kPa	Aeroshell undamaged, insulators cracked.
	Unit LRF-131 (perpendicular to shock front)		Dimetral strains, (diameter max./diameter min): 1.03
	Unit LRF-167 (parallel to shock front)		Dimetral strains, (diameter max./diameter min): 1.09
Flyer Plate Tests (Mound)	Bare, simulant-fueled capsules 20 mm dia., 3.5 mm. thick, 6061 Al Flyer plates (3 g)	1,100 m/s (plate) 330 m/s (capsule)	Significant deformation, no breaches
Capsule Projectile Test (U. of Dayton)	.50 caliber rifle barrel capsule in 2 piece Lexan sabot		
	152 x 152 x 12 mm cold rolled steel target		
	Plate angle 90° (Clad 043)	608 m/s	Capsule embedded in target, closure weld breached, vent pulvertized
	Plate angle 45° (Clad 039)	593 m/s	Capsule fragmented

Two LWRHU assemblies with open vents were aged for 1.75 y in 293 K seawater to tests seawater immersion behavior. One was placed in an aquarium at a 25-m depth (2.5 x 10^{-3} MPa pressure) for 640 days. The other unit was exposed to a pressure of 68.9 MPa (equivalent to a 6,000-m ocean depth) for 639 days. Weekly measurements of the plutonium concentration in the seawater were made. After the seawater exposure, the units were recovered and their graphite components were ashed. The amount of plutonium residue from the unit at 25-m depth was 32 ug and the 6000-m depth simulation was 154 ug. The conclusion from the data was that very fine particles of fuel were probably pumped through the vent frit and deposited in the graphite when the hydrostatic pressure was released. As long as the capsules remain submerged and the graphite remains intact, there is probably little risk of fuel release. The capsule materials and fuel pellets did not seem to react to the seawater.

Over-pressure tests were conducted on two units to tests behavior under estimated maximum over-pressures that might be experienced in a launch pad or post-launch explosion. A shock tube was used in these tests. The results of exposure to static over-pressure as high as 12.76 ± 0.69 MPa was no release of fuel. However, the graphite components were striped away. This left no added protection in subsequent collisions.

To test reaction of the LWRHU to burning propellant in case of a launch pad accident event, one unit was exposed to a 10.5 minute fire of burning solid rocket propellant. Propellant temperatures of about 2,336 K out to a distance of at least 1.8 m were measured. The tests showed that the aeroshell was intact, but the surface that had been exposed to the fire was eroded and encrusted with propellant fine products. The disassembled unit found the outer and middle pyrolytic graphite insulator bodies unchanged. The inner insulator body, however, had reacted with the Pt-Rh capsule, presumably forming a Pt/Rh-C eutectic. The platinum vent frit had disappeared and the capsule wall thickness was reduced 40% in places. In summary, the capsule integrity was greatly reduced by the fire exposure, but the outer graphite components provided sufficient containment capability to prevent fuel release.

Fragments could impact the LWRHU from a space shuttle external tank explosion accident. Simulations of the behavior to fragmentation events used a 0.5-caliber aluminum alloy bullets with velocities ranging from about 305 to 940 m / s. As summarized in Table 2, at velocities up to 757 m / s, the LWRHU was deformed but not breached. At higher velocities (908 and 940 m / s) no physical remnants of the capsules were found, but a significant amount of uranium was detected in the graphite debris. In conclusion, the LWRHU was not breached by collisions with compact, external tank fragments at velocities up to 757 m / s.

Impact tests simulated the behavior of the LWRHU of a terrestrial impact caused by an energetic launch or suborbital explosion that dissembles the spacecraft and removes some portion of the heat source protection.

Head-on, side, and at a 45° angle to the horizontal impact tests were performed at 48 m / s. The capsule in all cases showed some deformation, but no breaches. Additional impact tests were performed on side-on arrangements because this was determined to be the worst orientation. Velocities of 68, 88, 105 and 128 m / s were tested. Results were capsule deformation increased with velocity, but no rupture of the Pt-30%Rh clads were observed. Another two tests were conducted with a graphite reentry ablation shield. These showed considerably less deformation of the capsule. In summary, for impact velocities as high as 128 m / s (a test velocity derived from the Nuclear Regulatory Commission Containment Specification for air shipment of plutonium), no fuel release occurred from a bare LWRHU capsule.

In summary, safety analysis and tests validated that the LWRHU met the safety criteria for NASA missions.

Examining the use of LWRHUs in a spacecraft, Galileo used 117 units. Fig. 9 shows the distribution of the units on the spacecraft. The Galileo spacecraft was launch using the Space Shuttle. A probability and risk assessment together results in a Failure Abort Sequent Tree (FAST). The top-level assessment, with probabilities, is shown in Fig. 10 for each of the operation phases. The risk from a mission phase considers all the accidents in that phase as represented by the expectation case source term. The mission risk is the sum of the risk of all mission phases. In general, the Galileo Mission had a greater than 98% chance of success. Phase 5 is the only phase that has been identified with a small probability of a plutonia release from LWRHUs; 5 X 10^{-7} per mission.[5]

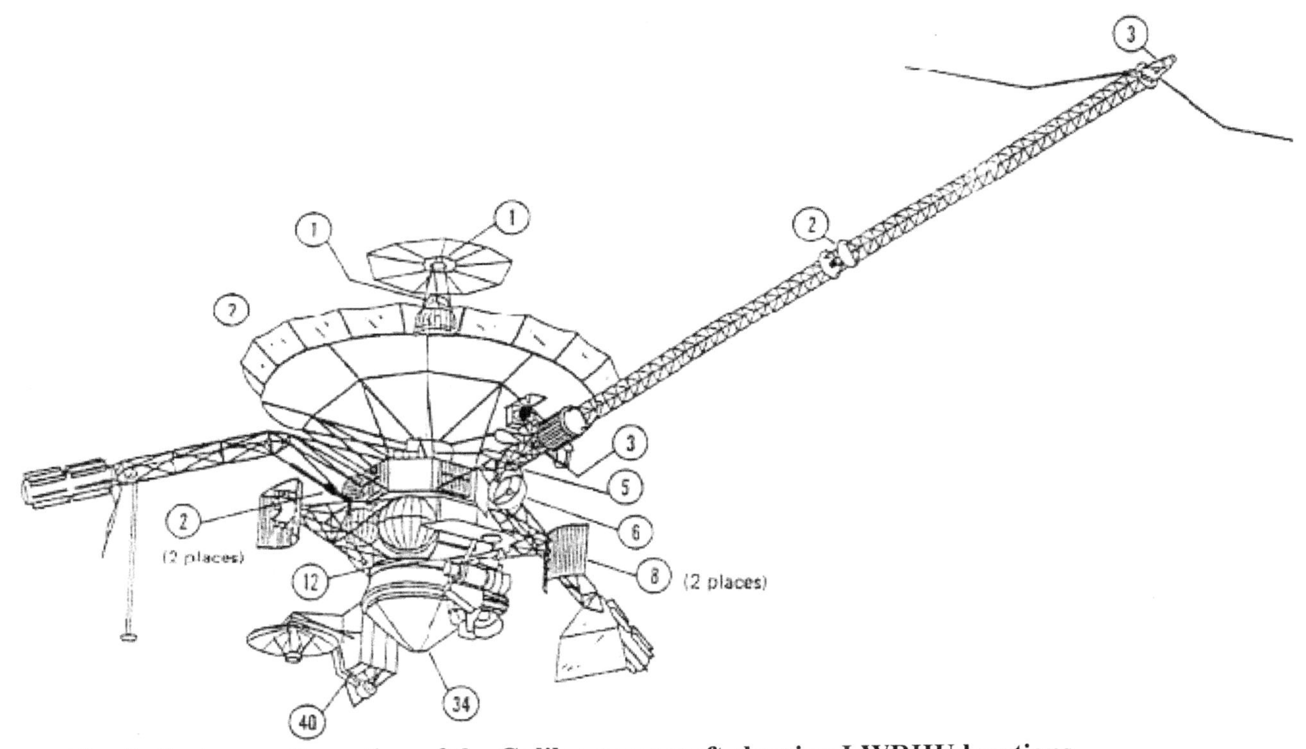

Fig. 9. Cruise configuration of the Galileo spacecraft showing LWRHU locations.

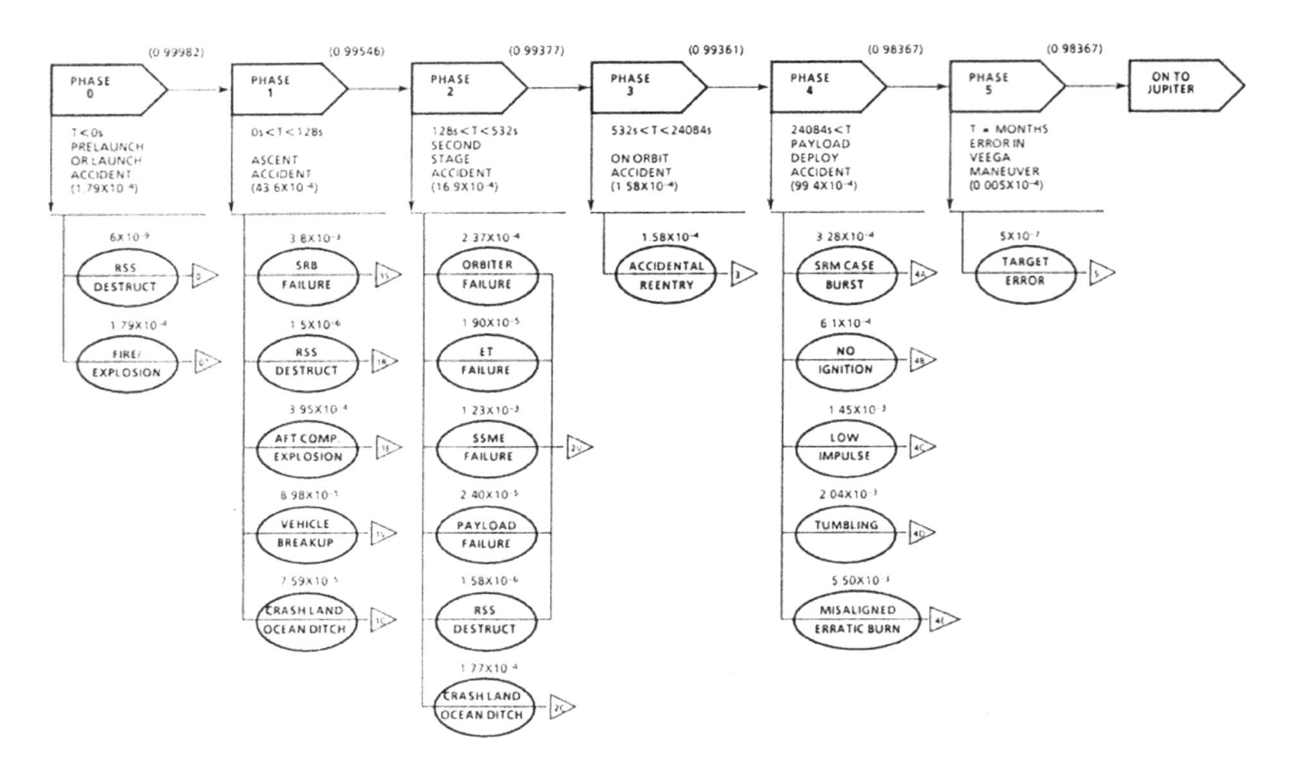

Fig. 10. The top-level LWRHU Failure Abort Sequence Tree (FAST).

Summary

Radioisotope Heater Units will continue to provide local heat sources when needed on space missions. Some 240 of these small, 1 watt units have been extremely successfully used on various missions. These included 117 LWRHUs used on the Galileo mission and 82 on the Cassini-Huygens spacecraft with the associated Huygens probe containing 35 units. The RHUs are light-weight, reliable units that have proven to be able to operate in very severe environments. Also, they operate independent of the solar flux. Extensive safety testing has validated that the designs meet all safety criteria.

Chapter 7

Potential Future Applications and Issues

Potential Future Applications

Fig. 1 illustrates the configuration of our solar system. Future applications of radioisotope systems are discussed in terms of: (1) science drivers and (2) manned exploration. Science drivers have been the focus to date. Even in the Apollo program, the Astronauts used the radioisotope generators for powering scientific equipment.

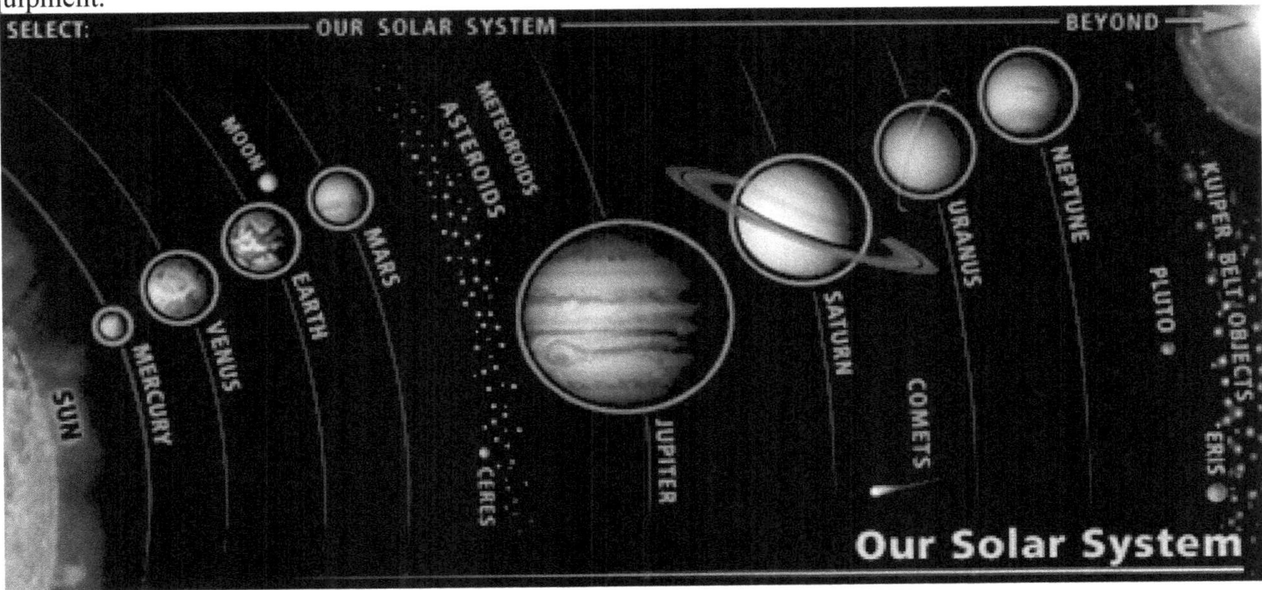

Fig. 1. Our solar system. ***Courtesy of NASA.***

Meeting the established scientific goals will require the current generation and more advanced radioisotope generators. NASA's goals are to answer the following questions:[1]

- How did the Sun's family of planets and minor bodies originate?
- How did the Solar System evolve to its current diverse state?
- What are the characteristics of the Solar System that led to the origin of life?
- How did life begin and evolve on Earth and has it evolved elsewhere in the Solar System?
- What are the hazards and resources in the Solar System environment that will affect the extension of human presence in space?

Also, for Mars exploration, the goals are:

- Determining if life ever arose on Mars;
- Understanding the process and and history of climate on Mars;
- Determining the evolution of the surface and interior of Mars; and
- Preparing for human exploration.

Programmatic considerations determine what mission will be performed and when. For Solar System exploration, there are three mission classes: namely Discovery (small), New Frontiers (NF) (medium), and Flagship (large). For Mars exploration, category names are Scout (small), medium, and large. Discovery and Scout missions have cost caps that effectively eliminate the consideration of radioisotope generators. Other classes, New Frontiers and Flagship, can consider the use of radioisotope generators. When adjusted for inflation, New Frontier missions have a cost cap of $767M (FY06). Flagship missions have cost caps of $750M to $3B. Cost caps for Mars programs are not specified, but are expected to be similar.

NASAs 2006 Solar System Exploration Roadmap[2] identified a set of missions for Solar System exploration out to 2035. These are shown in Fig. 2 and 3. Generally the reference mission set requires more funding than will be available. However, it provides a bases for establishing future radioisotope generator development and fabrication planning. The missions include planetary flybys, orbiters, sample and return, and rovers.

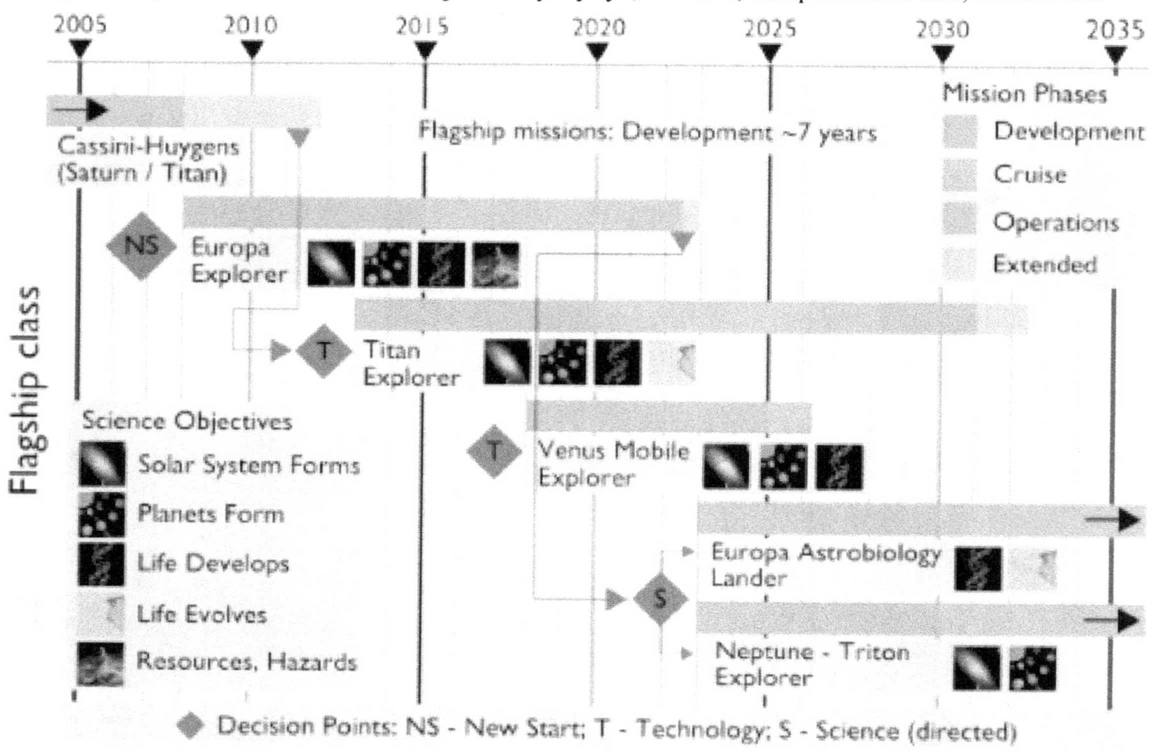

Fig. 2. Recommended sequence of Flagship Missions established by the SSE Roadmap Team.

The missions in Fig. 2 and 3 are a subset of a larger list of possible missions. The mission priorities tend to change with time, especially as new scientific results emerge. Table 1 provides a list of other missions that are under consideration that could benefit from developments in radioisotope generators.

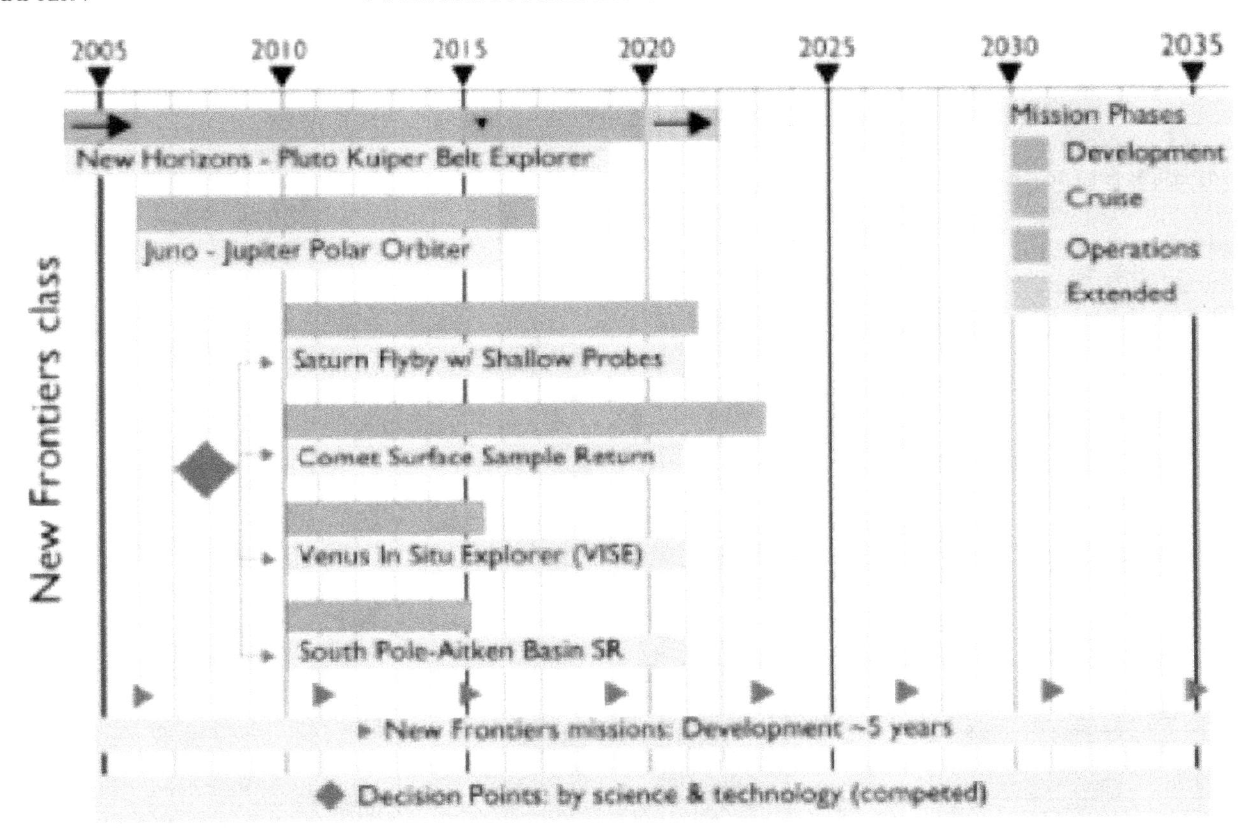

Fig. 3. Sequence of New Frontiers Missions established by the SSE Roadmap Team.

Currently, the GPHS-RTGs have been mainly used for planetary flyby and orbital missions. Solar flux decreases with the inverse square of the distance from the Sun. Since the solar panel mass and power scales linearly, and the power output reduces with the increasing distance from the Sun, at around 4 AU radioisotope power systems become more mass efficient than solar panels at comparable power outputs. Jupiter is 5.2 AU from the Sun; therefore, Jupiter and all planets further away from the Sun benefit from the use of radioisotope power systems (RPSs).

Mission environments have a significant impact on the type of radioisotope designs that will operate satisfactorily. Missions to the Jovian system encounter extreme radiation. The Galileo mission design minimized radiation exposure by using highly elliptic orbits to minimize the time spent in the high radiation environments. Orbiters and landers as proposed in the Europa Explorer, Europa Astrobiology Lander, Io Observer and Ganymede Observer missions, are continuously exposed to high radiation levels. The MMRTGs are a good candidate for these missions. The hundreds of redundant thermocouples means that the converter will continue to function even if a small number fail. The MMRTG can tolerate multi-MRad levels of radiation without additional shielding. On the other hand, Stirling Radioisotope Generators (SRGs) can interfere with science measurements from the EMI radiation. Also, SRGs are less forgiving in that they do not degrade gracefully.

Some mission require bursts of high power, such as for communications or traversing surfaces. RPSs provide a continuous power output. These power burst requirements could be accommodated by augmenting the RTPs with batteries. The batteries would be recharged from the RPSs during low power modes.

Table 1. Radioisotope Power System Design Reference Missions[a]

DRM Missions Name	Mission Class	Earliest Launch	Comments
Europa Explorer	Flagship	2015	MMRTGs, high radiation
Titaln Explorer (no Orbiter)	Flagship	2020	RPS excess heat for balloon heating
Titan Explorer (with Titan Orbiter)	Flagship	2020	Aerocapture, RPSs for orbiter in-situ
Venus Mobile Explorer	Flagship	2025	Special Stirling with active cooling
Europa Astrobiology Lander	Flagship	2030	EE follow on, high radiation
Neptune Triton Orbital Tour	Flagship	2030	RPS excess heat for Triton lander
Neptune Orbiter with Probes	Flagship	2030	Neptune aerocapture
Neptune Orbiter/Triton Explorer	Flagship	2030	RPS excess heat for Triton lander
Uranus Orbiter with Probes	Flagship	2035	RPS required, Galileo like configuration
Saturn Ring Observer	Flagship	2035	NRC DS recommended
Neptune Flyby	New Frontiers	2020	RPS required, New Horizons like
Uranus Flyby	New Frontiers	2020	RPS required, New Horizons like
Neptune Flyby with Probes	New Frontiers	2020	Jupiter or Saturn Entry Probes like
Uranus Flyby with Probes	New Frontiers	2020	Jupiter or Saturn Entry Probes like
Io Observer	New Frontiers	2020	Higher radiation than at Europa
Ganymede Observer	New Frontiers	2020	Lower radiation than at Europa
Enceladus Explorer	New Frontiers	2020	New mission, based on new finding
Trojan/Centaur Recon Flyby	New Frontiers	2020	REP--requires over 8 W/kg
Venus Geophysical Network	New Frontiers	2020	Special Venus RPS
Mercury Geophysical Network	New Frontiers	2020	RPS required at dark Polar Regions
Mars Science Laboratory	Large	2010	Baselined with one MMRTG
Mars Astrobiology Field Laboratory	Large	2016	Possibly MMRTG heritage
Mars Multi-Lander Network	Large	2020	Small RPS (baselined solar power)
Mars Mid-Rovers	Medium	2016	Small RPS (baselined solar power)

There is a desire to perform long-life, in-situ missions on Venus. Venus, though located only 0.72 A.U. from the Sun, has an extreme environment of temperatures (surface temperature is 735 K) and pressure (9,000 kPa). Currently, the Stirling Radioisotope Generator radiator operating at temperatures around 363 K would probably necessitate design modifications. Also, the corrosive atmosphere on Venus may require further design changes.

Lunar science laboratories and the support of crewed lunar surface operations can also benefit from RPS.

Table 2 provides some representative power requirements for some of the above potential missions.[3] The criticality level numbers for needing an advanced radioisotope power system are defined as follows:

1) Enabling - provides for achieving the science objectives of the mission with an affordable launch vehicle (Delta 3 class or smaller).

2) Strongly enhancing - provides substantial increase in payload or reduction in cost or risk.

3) Enhancing - provides increase in payload or reduction in cost or risk.

Table 2. Advanced RPS top level requirements for near-future missions.

Power (W_e)	Mass (kg)	Lifetime (Yrs.)	Efficiency (%)	Voltage (Vdc)	Potential Missions	Criticality Level
0,04 to 0.10	0.25	20	4 - 5	5	Mars Weather/Seismic Stations	1
1 to 2	0.5	5 - 10	5 - 10	5	Europa Lander Surface Laboratory	1
10 to 20	2	15 - 20	15	5	Surface In-situ Laboratories Aerobots or Aero-rover	2
50 to 100	7 to 10	4 to 5	18 to 20	28	Rover and Sample Retriver	2 or 3
100 to 200	8 to 20	10	18 to 21	28	Europa Lander	1
100 to 200	8 to 20	15	18 to 21	28	Titan Explorer	2
100 to 200	8 to 20	15	18 to 21	28	Neptune Orbiter	1
100 to 200	8 to 20	15	18 to 21	28	Saturn Ring Observer	1
100 to 200	8 to 20	15 to 30	18 to 21	28	Interstellar Probe	1

Issues and Trends

The characteristics of thermoelectric converters and dynamic converters must be considered in selecting candidates for future missions. Thermoelectric converters decay gracefully, have demonstrated in flight missions of tens of years of operation, and incorporate highly redundant elements. However, they suffer from relatively low efficiency, requiring more plutonium-238 fuel for a given power level, and low specific power. Dynamic converters, like the Stirling engine cycles, have not reached flight status, lack the inherent redundant features favored by NASA, have not demonstrated the tens of years lifetimes required for many missions, and may cause EMI radiation interference with the payload. They offer the benefit of much higher efficiencies, require much less plutonium-238 fuel for a given power level, and much improve specific power.

Plutonium-238 is in short supply and expensive. Hence, there is a major interest in higher efficient conversion systems. The higher efficient systems also tend to have improved specific power and extend the useful range of radioisotope power system from about 1 kWe to 10 kW_e. Fig. 3 illustrates the PuO_2 requirements as a function of power level and conversion efficiency.[4]

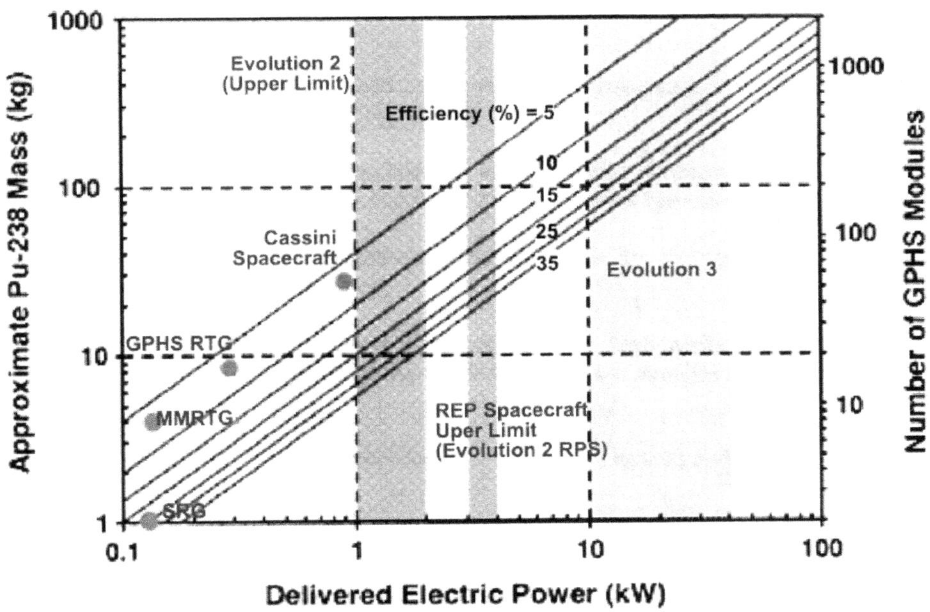

Fig. 3. Plutonium requirements versus delivered power.

A significant issue is how to insert new technologies into programs when spacecraft development teams are generally risk averse. Past development and continuing activities have made significant efforts to develop higher efficiency converters with dynamic system and improved thermoelectric materials. The dynamic systems have demonstrated engineering prototypes in Rankine, Brayton and Stirling converters. Currently, the Advanced Stirling Radioisotope Generator system is in flight system development with a goal of flight readiness by 2013.

Though ground demonstrations of new technologies are necessary, the ultimate test is in a space application. However, expensive missions try to avoid technologies that have not been proven in space. This creates a dilemma. In order to avoid initial use of the ASRG on expensive New Frontier or Discovery Class missions until flight proven, NASA is examining how to use them on a Discovery Class mission where it would enhance or enable the scientific return. The missions being examined include:[5]

- Lunar polar exploration,
- Moon rover/comsat,
- Titan Mare explorer,
- Io fly-by volcano observer,
- Trojan asteroid mission lander,
- Comet Hopper lander,
- Comet sample and return,
- Mars lander drill, and
- Venus balloons.

These studies have resulted in NASA offering up to two ASRGs as government furnished equipment within a Discovery Announcement of Opportunity.

Fig. 4 illustrates the expected improvements underway beyond the GPHS-RTG. The MMRTG provides the advantage over the GPHS-RTG of being able to operate in the Marian environment. ASRG provides the additional advantages of using many less GPHS modules for a given power level and having a 37% gain in specific power over the GPHS-RTG for the 650 C hot end temperature design; the 850 C design shows a 57% gain.

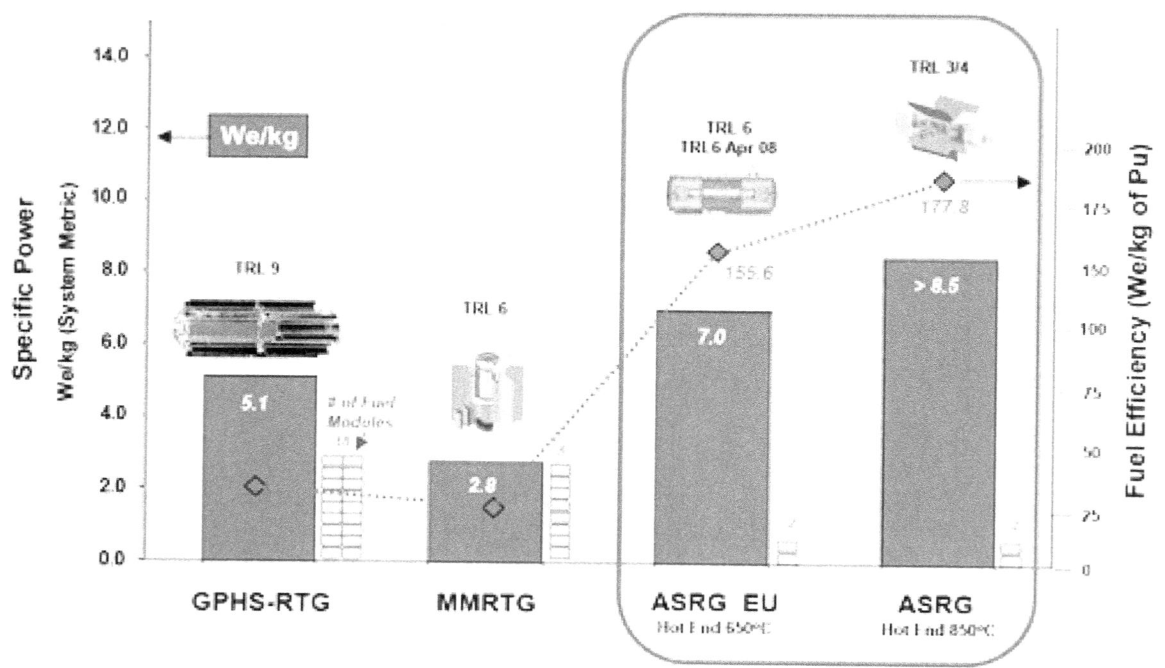

Fig. 4. ASRG provides a significant improvement in radioisotope capability. ***Courtesy of NASA***

Both the Multi-Mission Radioisotope Thermoelectric Generator (MMRTG) and the Advanced Stirling Radioisotope Generator (ASRG) continue to use the General Purpose Heat Source (GPHS) modules for their heat source. The first MMRTG is planned to be used for the Mars Science Laboratory mission scheduled for launch in 2011. The MMRTG specific mass (2.8 W_e / kg) is lower than that of the GPHS-RTG. However, it is capable of operating on the surface of Mars, the moon and in deep space. At 120 watts of electrical power, it can also be used on early robotic lunar science missions, particularly those involving long-duration operations at higher latitudes and in permanently shadowed regions. The ASRG can also perform the same missions as the MMRTG once it is space qualified. A major advantage of the ASRG is its high efficiency reduces the need for Pu-238 by a factor of four for a given electric power level.[6]

Advanced thermoelectric thermocouples beyond SiGe have failed to attain flight status. It is a matter of time and again maintaining a sustained effort whether the current candidates will reach flight readiness. These include the candidate materials: p-type Zintl ($Yb_{14}MnSb_{11}$), n-type Lanthanum Telluride ($La_{3-x}Te_4$), a nano-structure p-type SiGe and a mechanically alloyed n-type SiGe.

A following phase of radioisotope generator development is expected to lead to the development of higher powered radioisotope generators to support precursor and crewed lunar surface operations and more ambitious applications on robotic planetary missions. Power levels could be in the 1 to 2 kW_e range. These power level could possibly support radioisotope electric propulsion for planetary missions and highly capable robotic rovers, crewed rovers, habitat power, and autonomous science stations for lunar applications. The heat source would probably continue to use the GPHS; improvements would be in the power conversion technology.

In the future, surface power unit on the order of several tens of kilowatts will eventually be required. These unit would be designed specifically to meet the needs of highly advanced robotic missions, in-situ resource utilization missions, extended crewed stays on the lunar surface, and eventual missions to Mars. Though a fission power source is a more likely candidate, multiple GPHS modules could be advantageous from a lifetime and reliability standpoint and the reduce need for radiation shielding.

NOTES

OVERVIEW

[1] Dennis Miotla, “Assessment of Plutonium-238 Production Alternatives,” Briefing for Nuclear Energy Advisory Committee, April 21, 2008.
[2] H, M. Dieckamp, *Nuclear Space Power Systems,* Atomics International, Canoga Park, CA, September 1967
[3] Atomic Industrial Forum, *Guidebook for the Application of Space Nuclear Power Systems,* New York, January 1969.
[4] D. Buden et al., "Selection of Power Plant Elements for Future Reactor Space Electric Power Systems," LA-7858, Los Alamos National Laboratory, NM, September 1979.
[5] Ibid, D. Buden, et al.
[6] U.S. Department of Energy, "Environmental Development Plan (EDP)--Space Applications," DOE/EDP-0026, April 1978.
[7] “Nuclear Energy in Space,” DOE Office of Nuclear Energy, Science & Technology, DOE/NE0071.
[8] Bennett, Gary, et al, “Mission of Daring: The General-Purpose Heat Source Radioisotope Thermoelectric Generator,” 4th International Energy Conversion Engineering Conference, AIAA 2006-4096, June 2006.
[9] Jet Propulsion Laboratory Internet Site, www.jpl.nasa.gov.
[10] U. of Wisconsin lectures, fti.neep.wisc.edu/neep602/SPRING00/lecture34.pdf
[11] U.S. Department of Energy, ebid.

CHAPTER ONE

[1] G. L. Bennett, J. J. Lombardo, and B. J. Rock, “Space Nuclear Electric Power System,” *Advances In Astronautical Sciences, Volume 44*, American Astronautical Society, 1980.
[2] “Nuclear Energy in Space,” DOE Office of Nuclear Energy, Science & Technology, DOE-0071 NE.
[3] G. Bennett, et al., “Mission of Daring: The General-Purpose Heat Source Radioisotope Thermoelectric Generator,” Fourth International Energy Conversion Engineering Conference, AIAA 2006-4096, June 2006.
[4] Jet Propulsion Laboratory internet site, www.jpl.nasa.gov.
[5] Harmon, B. Alan and David B. Lavery, “NASA Radioisotope Power Systems Program Update,” *CP969, Space Technology and Applications International Forium--STAIF 2008*, edited by M. S. El-Genk, 2008 American Institute of Physics, 978-0-7354-0486-1/08, pp. 396-402.
[6] John C. Mankins, “Technology Readiness Levels,” White Paper, April 6, 1995, Office of Space Access and Technology, NASA.
[7] Gary L. Bennett, “A Look At The Soviet Space Nuclear Power Program,” The 24th Intersociety Energy Conversion Engineering Conference, Washington, D. C., 6 - 11 August 1989.

CHAPTER TWO

[1] Robert G. Lange and Edward F. Mastal, “A Tutorial Review of Radioisotope Power Systems,” in the book *A Critical Review of Space Nuclear Power and Propulsion 1984-1993*, Edited by Mohamed S. El-Genk, American Institute of Physics Press, New York, ISBN 1-56396-317-5, pp. 1-20, 1994,
[2] Gary L. Bennett, “Lessons Learned From The Galileo and Ulysses Flight Safety Review Experience,” *CP420, Space Technology and Applications International Forum-1998*, edited by Mohamed S. El-Genk, 1998, pp. 1269 - 1274.
[3] Gary L. Bennett, et al., “Development and Implementation of a Space Nuclear Safety Program,” *Space Nuclear Power Systems, 1987*, Edited by M. El-Genk and M. D. Hoover, Orbit Book Co., Malabar, FL, 1988, pp. 59 - 92.

[4] Jerry L. Wert, Dennis L. Oberg, and Tommy L. Criswell, "Effect of Radiation from an RTG on the Installation, Personnel, and Electronics of a Launch System," *Space Nuclear Power Systems, 1989*, Edited by M. El-Genk and M. D. Hoover, Orbit Book Co., Melbourne, Fl, 1992, pp. 151 - 157.
[5] NASA Document, "Final Environmental Impact Statement for the New Horizons Mission," July 2005.
[6] Lisa Herrera and Beverly A. Cook, "Support Facilities For Radioisotope Applications In Space Environments," *NTSE-92 Nuclear Technologies For Space Exploration*, American Nuclear Society Meeting, Jackson, Wyoming, August 16 - 19, 1992, pp. 114 - 122.
[7] "DOE Environmental Impact Statement for the Proposed Consolidation of Nuclear Operations Related to Production of Radioisotope Power Systems," Office of Nuclear Energy, Updated 11/4/2006.
[8] Dennis Miotla, "Assessment of Plutonium-238 Production Alternatives," Briefing for Nuclear Energy Advisory Committee, April 21, 2008.
[9] Brian Berger, "Disagreement Arises About Plutonium-238 Storage," Space News, April 7, 2008, page 6.

CHAPTER THREE

[1] Joseph A. Angelo, Jr. and David Buden, *Space Nuclear Power*, Orbit Book Company, 1985.
[2] Robert G. Lange and Edward F. Mastal, "A Tutorial Review of Radioisotope Power Systems," in *A Critical Review of Space Nuclear Power and Propulsion, 1984-1993*, Editor Mohamed S. El-Genk, 1994 American Institute of Physics, 1994, pp 1 - 20.
[3] H. Bateman, *Cambridge Philosophical Society Proceedings*, 15, pp. 423-427, 1910.
[4] D. Buden, et al, "Selection of Power Plant Elements for Future Reactor Space Electric Power Systems," LA-7858, Los Alamos National Laboratory, September 1979.
[5] R. V. Anderson, et al, "Space Reactor Electric Systems," ESG-DOE-13398, Rockwell International, 29 March 1983.
[6] Ibid D. Buden, et al.
[7] C. Wood, "Optimization of Thermoelectric Materials," *Proceedings of 1982 Working Group on Thermoelectrics*, 7-9 Dec 1982, Jet Propulsion Laboratory, JPL-D-497, January 1983.
[8] NASA, "Space Shuttle Program-Space Shuttle System Payload Accommodations, Level II Program Definition and Requirements, Vol. XIV," JSC-07700, Rev H, 16 May 1983.
[9] NASA, *Space and Planetary Environment Criteria Guidelines for Use in Space Vehicle Development*, 1982 Revision (Volume I), NASA TM-82478, January 1983.
[10] F. Kreith, *Radiation Heat Transfer for Spacecraft and Solar Power Plant Design*, International Textbook Co., Scranton, 1962.
[11] John Dassoulas and Ralph L. McNutt, Jr., "RTGs on Transit," *CP880, Space Technology and Applications International Forum--STAIF 2007*, edited by M. S. El-Genk, 2007 American Institute of Physics 978-0-7354-4/07, pp. 195 - 204
[12] G. R. Grove, J. A. Powers, N. Goldenberg, D. P. Kelly, and D. L. Prosser, "Plutonium-238 Isotopic Heat Sources: A Summary Report," Mound Laboratory Report MLM-1270, Miamisburg, Ohio, 7 June 1965.
[13] G. L. Bennett, J. J. Lombardo, and B. J. Rock, "U.S. Radioisotope Thermoelectric Generator Space Operating Experience (June 1961-December 1982)," *18th Intersociety Energy Conversion Engineering Conference*, Orlando, FL, 21-26 August 1983.
[14] P. Ronklove and V. Truscello, "Long-Term Tests of a SNAP-19 Thermoelectric Generator," *Intersociety Energy Conversion Engineering Conference Proceedings, 1972*, pp. 186-193.
[15] C. J. Goebel and L. R. Putnam, "SNAP-19 Performance Update for Pioneer and Viking Missions," *14th Intersociety Energy Conversion Engineering Conference*, 5-10 August 1979, pp. 1476-1479
[16] J. A. Van Allen, "Pioneer's Unfunded Reach for the Stars," *Aviation Week & Space Technology, 12 April 1982*, p. II
[17] Wayne M. Brittain, "SNAP-19 Radioisotope Thermoelectric Generator Thermal/Electrical Integration With the Viking Mars Lander," *Intersociety Energy Conversion Engineering Conference Proceedings, 1971*, pp. 1189-1199.

[18] G. Stapfer and V. Truscello, "The Long-Term Behavior of SNAP-19 Generators Which Use the TAGS Thermoelectric Material," *Intersociety Energy Conversion Engineering Conference Proceedings, 1971*, pp. 1184-1189.
[19] B. M. French and S. P. Maran, "A Meeting With the Universe," *NASA, EP-177*, 1981
[20] Ibid French and Maran 1981.
[21] "Viking '75, SNAP/Viking Lander System, Final Safety Analysis Report, Vol. I," *Teledyne Isotopes Report ESD-3069-15-1*, Timonium, Md.
[22] "SNAP-19/Pioneer F Safety Analysis Report, Vol. I, Reference Design Document," *Teledyne Isotopes, INSD-2873-42-1*, Timonium, Md., December 1970.
[23] A. A. Pitrolo, B. J. Rock, W. Remini, and J. A. Leonard, "SNAP-27 Program Review," *Intersociety Energy Conversion Engineering Conference Proceedings*, 1969, pp. 153-170.
[24] W. C. Remini and J. H. Grayson, "SNAP-27 / ALSEP Power Subsystem Used in the Apollo Program," *Intersociety Energy Conversion Engineering Conference, 1970 Proceedings*, pp. 13- 10 to 13-15.
[25] "TRANSIT RTG Final Safety Analysis Report, Vol. I," *TRW report TRW(A)-11464-0491*, March 1971.
[26] H. Lurie and S. Rocklin, "Transit RTG-A Status Report," *Intersociety Energy Conversion Engineering Conference, 1970 Proceedings*, pp. 15-111 to 15-116.
[27] "Multi-Hundred Watt Radioisotope Thermoelectric Generator Program, Final Safety Analysis Report for the MJS-77 Mission," *General Electric Document No. 77SDS4206*, January 1977.
[28] H. Lurie and S. Rocklin, "Transit RTG-A Status Report," *Intersociety Energy Conversion Engineering Conference, 1970 Proceedings*, pp. 15-111 to 15-116.
[29] "Multi-Hundred Watt Radioisotope Thermoelectric Generator Program, Final Safety Analysis Report for the MJS-77 Mission," *General Electric Document No. 77SDS4206*, January 1977.
[30] "Multi-Hundred Watt Radioisotope Thermoelectric Generator Program Final Safety Analysis Report," LES 8/9, General Electric Document No. GEMS-419, March 1975.
[31] "Power for Space Experimentation," *General Electric Pamphlet GEA-9326(8-74)IM.*
[32] L. Garvey and G. Stapfer, "Performance Testing of Thermoelectric Generators Including Voyager and LES 8/9 Flight Results," *Intersociety Energy Conversion Engineering Conference, 1979* Proceedings, pp. 1470-1475.
[33] C. E. Kelly, "The MHW Converter (RTG)," *Intersociety Energy Conversion Engineering Conference, 1975 Proceedings*, pp. 880- 886.
[34] V. Shields, "Long-Term Behavior of Silicon Germanium Thermoelectric Generators," *Intersociety Energy Conversion Engineering Conference, 1981* Proceedings, pp. 315-320
[35] Ibid Kelly 1975.
[36] V. Shields, "Long-Term Behavior of Silicon Germanium Thermoelectric Generators," *Intersociety Energy Conversion Engineering Conference, 1981* Proceedings, pp. 315-320.

CHAPTER FOUR

[1] E. C. Snow, R. W. Zocher, I. M. Grinberg, and L. E. Hulbert, "General-Purpose Heat Source Development Phase 11-Conceptual Designs," Los Alamos National Laboratory Report LA-7546-SR, November 1978.
[2] Ibid E. C. Snow, R. W. Zocher, I. M. Grinberg, and L. E. Hulbert,1978.
[3] A. Schock, "Design, Evolution and Verification of the General Purpose Heat Source," Intersociety Energy Conversion Engineering Conference, 1980 Proceedings, pp. 1032-1042.
[4] G. L. Bennett, J. L. Lombardo, and B. J. Rock, "Nuclear Electric Power for Space Systems: Technology Background and Flight Systems Program," Intersociety Energy Conversion Engineering Conference, 1981 Proceedings, pp. 362-368.
[5] Gary Bennett, et al, "Mission of Daring: The General Purpose Heat Source Radioisotope Thermoelectric Generator," 4th International Conversion Engineering Conference, AAIA 2006-4096, 26-29 June 2006.
[6] Robert G. Lange and Edward F. Mastal, "A Tutorial Review of Radioisotope Power Systems," in *A Critical Review of Space Nuclear Power and Propulsion 1984-1993*, Edited by Mohamed S. El-Genk, American Institute of Physics, L.C. Catalog Card No. 94-70780, 1994, pp. 1 - 20.

[7] Ibid Gary Bennett, et al, 2006.
[8] Ibid G. L. Bennett, J. L. Lombardo, and B. J. Rock, 1981
[9] Ibid Gary Bennett, et al, 2006.
[10] Ibid Gary Bennett, et al, 2006
[11] Ibid Gary Bennett, et al, 2006.
[12] Gary L. Bennett, et al, "Development and Implementation of a Space Nuclear Safety Program," *Space Nuclear Power Systems*, 1987, Edited by M. S. El-Genk and M. D. Hoover, Orbit Book Co., Malabar, FL, 1988, pp. 59 - 92.
[13] Gary L. Bennett, et al, "Update to the Safety Program for the General Purpose Heat Source Radioisotope Thermoelectric Generators for the Galileo and Ulysses Missions," *Space Nuclear Power Systems* 1989, Edited by M. S. El-Genk and M. D. Hoover, Orbit Book Co. Malabar, FL., 1992, pp. 199 - 222.
[14] Gary L. Bennett, "Safety Aspects of Thermoelectrics in Space," published in CRC *Handbook of Thermoelectrics*, Edited by D. M. Rowe, CRC Press, New York, 1995, pp. 551 - 572.
[15] Marshall B. Eck and Meera Mukunda, "On the Nature of the Response of the General Purpose Heat Source to the Impact of Large Solid rocket Motor Casing Fragments," *Space Nuclear Power Systems* 1989, Edited by M. S. El-Genk and M. D. Hoover, Orbit Book Co., Malabar, FL., 1992, pp. 223 - 238.
[16] NASA Document, "Final Environmental Impact Statement for the New Horizons Mission," July, 2005.
[17] Alan V. Von Arx, "MMRTG Heat Rejection Summary," CP813, *Space Technology and Applications International Forum--STAIF 2006*, edited by M. S. El-Genk, 2006 American Institute of Physics 0-7354-0305-8/06, 2006, pp. 743-750.
[18] National Aeronautics and Space Administration, "Final Environmental Impact Statement for the Mars Science Laboratory Mission," Volume 1, November 2006.

CHAPTER FIVE

[1] "Nuclear Powered Organic Rankine Systems for Space Applications," *Proceedings of the AFOSR Special Conference on Prime Power for High Energy Space Systems*, Norfolk, Va. 22-25 February 1982.
[2] "Dynamic Isotope Power System Applications Study Results," General Electric Co., Space Division, April 1979.
[3] G. L. Sorensen, R. E. Niggemann, E. C. Krueger, R. C. Brouns, and F. A. Russo, "Status Report of the Dynamic Isotope Power System," *Intersociety Energy Conversion Engineering Conference, 1979 Proceedings*, pp. 1396-1400.
[4] "Nuclear Powered Organic Rankine Systems for Space Applications," *Proceedings of the AFOSR Special Conference on Prime Power for High Energy Space Systems*, Norfolk, Va. 22-25 February 1982.
[5] Gary L. Bennett and James J. Lombardo, "The Dynamic Isotope Power System: Technology Status and Demonstration Program," Chapter 20, *Space Nuclear Power Systems 1988*, Edited by M. El-Genk and M. D. Hoover, Orbit Book Company, Malabar, FL, 1989, pp. 101-115.
[6] Thomas J. Sutliff and Leonard A.Dudzinski, "NASA Radioisotope Power System Program--Technology and Flight Systems," 7th International Energy Conversion Engineering Conference, AIAA 2009-4575, August 2009, Denver, Colorado.
[7] Joseph A. Angelo, Jr and David Buden, *Space Nuclear Power*, Orbit Book Company, Malabar, Florida, 1985.
[8] J. P. Holman, *Thermodynamics* (2nd Edition), McGraw-Hill Book Company, New York, 1974.
[9] W. Reynolds and H. Perkins, *Engineering Thermodynamics,* McGraw-Hill Book Company, 1974.
[10] K. Wark, *Thermodynamics* (3rd Edition), McGraw-Hill Book Company, New York, 1977.
[11] I. Asimov, *Biographical Encyclopedia of Science and Technology* (Revised Edition), Doubleday and Company, New York, 1972.
[12] Lee S. Mason and Jeffrey G. Schriber, "A Historical Review of Brayton and Stirling Power Conversion Technologies for Space Applications," *NASA/TM--2007-214976*, November 2007.
[13] Ibid Gary L. Bennett and James J. Lombardo,1989.

[14] Pearson, Richard J., "Dynamic Isotope Power System Critical Component Verification,"*Space Nuclear Power Systems 1988*, Editors Mohamed S. El-Genk and Mark D. Hoover,Orbit Book Co., ISSN 1-41-2824, 1989, pp. 117-125.
[15] Overholt, David M., "Dynamic Power Conversion Systems For Space Nuclear Power--Section III: Brayton Cycle," Published in *A Critical Review of Space Nuclear Power and Propulsion 1984-1993*, Edited by Mohamed S. El-Genk, American Institute of Physics, L. C. Card No. 94-70780, 1994, pp. 358-379.
[16] Ibid Lee S. Mason and Jeffery G. Schreiber, 2007.
[17] Lee S. Mason, et al., "Status of Brayton Cycle Power Conversion Development at NASA GRC," *NASA/TM--2002-211304.*
[18] W. R. Determan and R. B. Harty, "DIPS Spacecraft Integration Issues," *Sixth Symposium on Space Nuclear Power Systems*," Institute for Space Nuclear Power Studies, CONF-890103--Summs., Albuquerque, NM, 1989, pp. 181-183.
[19] William D. Otting, et al., "Dynamic Isotope Power System, Integrated System Test," *CONF 940101, 1994 American Institute of Physics*, 1994, pp. 365 - 369.
[20] Richard B. Harty, "Comparison Of DIPS And RFCs For Lunar Mobile And Remote Power Systems," *Space Nuclear Power Systems Ninth Symposium*, Editors Mohamed S. El-Genk and Mark D. Hoover, AIP Conference Proceedings 246, L.C. Catalog Card No. 91-58793, 1992, pp. 202 - 207.
[21] G. L. Bennett, J. J. Lombardo, and B. J. Rock, "Space Nuclear Electric Power Systems," *Advances in Astronautical Sciences, Volume 44*, American Astronautical Society, 1980.
[22] G. L. Sorensen, R. E. Niggemann, E. C. Krueger, R. C. Brouns, and F. A. Russo, "Status Report of the Dynamic Isotope Power System," *Intersociety Energy Conversion Engineering Conference, 1979 Proceedings*, pp. 1396-1400.
[23] Ibid G. L. Bennett, J. J. Lombardo, and B. J. Rock, 1980.
[24] Sundstrand Corporation, "Final Report, Technology Verification Phase, Dynamic Isotope Power System," *Report No. 2032*, Sunstrand Energy Systems, Rockford, IL, 1982.
[25] Johnson, R. A., "SNAP-DYN Proven Power For Space," *Space Nuclear Power Systems Vol. VIII,* Editors Mohamed S. El-Genk and Mark D. Hoover,Orbit Book Co., ISSN 1-41-2824, 1989, pp. 127-137.
[26] Ibid Joseph A. Angelo, Jr. and David Buden
[27] B. Goldwater, "Study of a Free Piston Stirling Engine-Linear Power Conversion System for a 10.5 kWe and 51.0 kWe Output Power," Mechanical Technology Inc., November 1975
[28] G. Walker, *Stirling Cycle Machines*, Oxford University Press, Ely House, 1973.
[29] W. Beale et ai., "Free Piston Stirling Engines- A Progress Report," SAE Paper 730647, June 1973.
[30] Krause, David L. and Pete T. Kantzos, "Accelerated Life Structural Benchmark Testing for a Stirling Convertor Heater Head," *CP813, Space Technology and Applications International Forum--STAIF 2006*, edited by M. S. El-Genk, 2006 American Institute of Physics 0-7354-0305-8/06, pp. 623-630.
[31] Wood, J. Gary, et al, "Advanced Stirling Convertor Update," *CP813, Space Technology and Applications International Forum--STAIF* 2006, edited by M. S. El-Genk, 2006 American Institute of Physics 0-7354-0305-8/06, pp. 640-652.
[32] Wood, J. Gary, et al, "Continued Development of the Advanced Stirling Convertor (ASC)", *Sunpower Report Doc. 0104* issued by AIAA.
[33] Jack Chan, J. Gary Wood, Jeffrey G. Schreiber, "Development of Advanced Stirling Radioisotope Generator for Space Exploration," *CP880, Space Technology and Applications International Forum--STAIF*, edited by M. S. El-Genk, 2007 American Institute of Physics 978-0-7354-0386-4/07, pp. 615 - 623.
[34] Ibid Wood, J. Gary, et al.
[35] Ibid Wood, J. Gary, et al.
[36] NASA, "Advanced Stirling Radioisotope Generator for NASA Space Science and Exploration Missions," M-1929, June 2007.
[37] Ibid, Thomas J. Sutliff and Leonard A.Dudzinski
[38] Ibid Harmon, B. Alan and David B. Lavery, 2008.
[39] T. Caillat, et al, "Advanced Thermoelectric Power Generation Technology Development at JPL," 3rd European conference on Thermoelectrics, Nancy, France, Sept. 2005.

[40] Andrew Mays, et al, "Lanthanum Telluride: Mechanochemical Synthesis of a Refractory Thermoelectric Material, " iCP969, *Space Technology and Applications International Forum*--STAIF, 2008, pp.672-678.
[41] Ibid Andrew Mays, et al, 2008.
[42] Cliff A. Spence, Michael Schuller, Tom R. Lalk, "Development, Evaluation, and Design Applications of an AMTEC Converter Model," *CP654, Space Technology and Applications Forum--STAIF 2003*, edited by M. S. El-Genk, 2003 American Institute of Physics 0-7354-0114-4/03, pp. 685 - 707.
[43] M. A. Ryan, "The Alkali Metal Thermal-To-Electric converter For Solar System Exploration," Jet Propulsion Laboratory Report, mryan@jpl.nasa.gov.
[44] Jack F. Mondt, et al, "Alkali Metal Thermal to Electric Conveter (AMTEC) Technology Development for Potential Deep space Scientific Missions," Jet Propulsion Laboratory report.
[45] Joseph C. Giglio, Robert K. Sievers and Edward F. Mussi, "Update of the Design of the AMTEC Converter for Use in AMTEC Radioisotope Power Systems," *CP552, Space Technology and Applications International Forum-2001*, edited by M. S. El-Genk, 2001 American Institute of Physics 1-56396-980-710-7/01, pp. 1047 - 1054.
[46] Ibid Joseph C. Giglio, Robert K. Sievers and Edward F. Mussi, 2001
[47] R. M. Williams, et al, "Challenges Facing Successful Development of Long Life, High Temperature, High Efficiency/Power AMTECs for Space Applications," *CP552, Space Technology and Applications International Forum-2001*, edited by M. S. El-Genk, 2001 American Institute of Physics 1-56396-980-710-7/01, pp. 1055 - 1065.
[48] Robert D. Cockfield, "Radioisotope Power System Options for Future Planetary Missions," *CP552, Space Technology and Applications International Forum--2001*, edited by M. S. El-Genk, 2001 american Institute of Physics, 1-56396-980-7/01, pp. 740 - 746.
[49] Ibid Jack F. Mondt, et al, 2002.
[50] Jack F. Mondt, et al, "Advance Radioisotope Power System Technology Development For NASA Missions 2011 And Beyond," *6th European Space Power Conference*, Dorto, Portugal, 6 May 2002.
[51] B. Xu, et al., "Integrated Bandpass Filter Contacts for GaSb Thermophotovoltaic Cells," *CP746, Space Technology and Applications International Forum--STAIF2005*, edited by M. S. El-Genk, 2005 American Institute of Physics 0-7354-0230-2/05, 2005, pp. 615-622.
[52] Christopher J. Crowley, et al., "Thermophotovoltaic Converter Performance for Radioisotope Power Systems," *CP746, Space Technology and Applications International Forum--STAIF2005*, edited by M. S. El-Genk, 2005 a\American Institute of Physics 0-7354-0230-2/05, 2005, pp. 601-614.
[53] V. L. Teofilo, et al., "Thermophotovoltaic Energy Conversion for Space Applications," *CP813, Space Technology and Applications International Forum--STAIF 2006*, edited by M. S. El-Genk, 2006 American Institute of Physics 0-7354-0305-8/o6, pp. 552-559.
[54] Ibid Christopher J. Crowley, et al., 2005.
[55] John C. Mankins, "Technology Readiness Levels," NASA Advanced Concepts Office, A White Paper, April 6, 1995.

CHAPTER SIX

[1] *Nuclear Power in Space,* DOE/NE-0071, US Department of Energy, Washington, D.C. 20585.
[2] Timothy G. George and Theresa A. Cull, "The Light-Weight Radioisotope Heater Unit (LWRHU): Development and Application," *Space Nuclear Power Systems 1987*, Edited by M. S. El-Genk and M. D. Hoover, Orbit Book Co., Malabar, FL, 1988, pp. 577 - 586.
[3] Gary H. Rinehart, "Light Weight Radioisotope Heater Unit (LWRHU) Production For The Cassini Mission," *CONF 970115, 1997 American Institute of Physics*, pp. 1375 - 1379.
[4] Ernest W. Johnson, "Safety Analysis For The Galileo Light-Weight Radioisotope Heater Unit," *Proceedings of the Seventh Symposium on Space Nuclear Power Systems, Institute For Space Nuclear Power Studies*, U. of New Mexico, 1990, pp. 581 - 588.
[5] Ibid Ernest W. Johnson, 1990.

CHAPTER SEVEN

[1] Tibor S. Balint, "NASA's RPS Design Reference Mission Set for Solar System Exploration," *CP 880, Space Technology and Applications International Form--STAIF 2007*, edited by M. S. El-Genk, 2007 American Institute of Physics 978-0-7354-0386-4/07, pp. 631 - 639.
[2] NASA--SSE Roadmap Team, "Solar System Exploration -- Solar system Exploration Roadmap for NASA's Science Mission Directorate, *Report Number: JPL-D-35618, NASA Science Missions Directorate, Planetary Science Division*, Washington, DC, 2006.
[3] Jack F. Mondt and Bill J. Nesmith, "Future Planetary Missions Potentially Requiring Radioisotope Power Systems," *CP504, Space Technology and Applications International Forum--2000*, edited by M. S. El-Genk, 2000 American Institute of Physics, 1-56396-919-X/00, pp. 1169 - 1174.
[4] George R. Schmidt and Michael G. Houts, "Radioisotope-Based Nuclear Power Strategy for Exploration Systems Development," *CP813, space Technology and Applications International forum--STAIF 2006*, edited by M. S. El-Genk, 2006 American Institute of Physics 0-7354-0305-8/06, pp. 334 - 339.
[5] Thomas J. Sutliff and Leonard A. Dudzinski, "NASA Radioisotope Power System Program--Technology and Flight Systems, " AIAA Paper 2009-4575, 7th International Energy Conversion Engineering conference, Denver, Colorado, August 2009.
[6] Ibid George R. Schmidt and Michael G. Houts, 2006.

CPSIA information can be obtained
at www.ICGtesting.com
Printed in the USA
LVHW061531280920
667301LV00031B/1992